Scientifically Speaking

A Dictionary of Quotations

About the Compilers

Carl C Gaither was born on 3 June 1944 in San Antonio, Texas. He has conducted research work for the Texas Department of Corrections, the Louisiana Department of Corrections, and taught mathematics, probability, and statistics at McNeese State University and Troy State University at Dothan. Additionally he worked for ten years as an Operations Research Analyst. He received his undergraduate degree (Psychology) from the University of Hawaii and has graduate degrees from McNeese State University (Psychology), North East Louisiana University (Criminal Justice), and the University of Southwestern Louisiana (Mathematical Statistics).

Alma E Cavazos-Gaither was born on 6 January 1955 in San Juan, Texas. San Juan has the name of a big city but in Texas it's just a small border town. She has previously worked in quality control, material control, and as a bilingual data collector. In addition to compiling the quotations for this science quotation book series she is also an SK1 in the United States Navy Reserve. She received her associate degree (Telecommunications) from Central Texas College and presently is working toward a BA degree with a major in Spanish and a minor in Art.

Together they selected and arranged quotations for the books *Statistically Speaking: A Dictionary of Quotations* (Institute of Physics Publishing, 1996), *Physically Speaking: A Dictionary of Quotations on Physics and Astronomy* (Institute of Physics Publishing, 1997), *Mathematically Speaking: A Dictionary of Quotations* (Institute of Physics Publishing, 1998), *Practically Speaking: A Dictionary of Quotations on Engineering, Technology and Architecture* (Institute of Physics Publishing, 1999) and *Medically Speaking: A Dictionary of Quotations on Dentistry, Medicine and Nursing* (Institute of Physics Publishing, 1999).

About the Illustrator

Andrew Slocombe was born in Bristol in 1955. He spent four years of his life at Art College where he attained his Honours Degree (Graphic Design). Since then he has tried to see the funny side to everything and considers that seeing the funny side to science has tested him to the full! He would like to thank Carl and Alma for the challenge!

Scientifically Speaking
A Dictionary of Quotations

Selected and Arranged by

Carl C Gaither
and
Alma E Cavazos-Gaither

Illustrated by Andrew Slocombe

Institute of Physics Publishing
Bristol and Philadelphia

British Library Cataloguing-in-Publication Data
A catalogue record for this book is available from the British Library.

ISBN 0 7503 0636 X

Library of Congress Cataloging-in-Publication Data are available

Production Editor: Simon Laurenson
Production Control: Sarah Plenty
Commissioning Editor: Jim Revill
Cover Design: Victoria Le Billon
Marketing Executive: Colin Fenton

Published by Institute of Physics Publishing, wholly owned by The Institute of Physics, London

Institute of Physics Publishing, Dirac House, Temple Back, Bristol BS1 6BE, UK
US Office: Institute of Physics Publishing, Suite 1035, The Public Ledger Building, 150 South Independence Mall West, Philadelphia, PA 19106, USA

Typeset in TEX using the IOP Bookmaker Macros
Printed in Great Britain by J W Arrowsmith Ltd, Bristol

We respectfully dedicate this book to
Brenda and Stan Burt and family
Mike and Vickie Gould and family

In memeory of
Clifford C Gaither
(March 1, 1917–February 8, 2000)
Husband, Father, and Friend
You will be missed

I vow to strive to apply my professional skills only to projects which, after conscientious examination, I believe to contribute to the goal of coexistence of all human beings in peace, human dignity and self-fulfillment.

I believe that this goal requires the provision of an adequate supply of the necessities of life (good food, air, water, clothing and housing, access to natural and man-made beauty), education, and opportunities to enable each person to work out for himself his life objectives and to develop creativeness and skill in the use of hands as well as head.

I vow to struggle through my work to minimise danger; noise; strain or invasion of privacy of the individual; pollution of earth, air or water; destruction of natural beauty, mineral resources and wildlife.

Thring, Meredith
New Scientist
Scientist's Oath
January 7, 1971

CONTENTS

PREFACE

The wisdom of the wise and the experience of ages, may be preserved by quotations.

Isaac Disraeli
Curiosities of Literature

The importance of reading books that have become classics is not just to be able to know the level of understanding in olden times—a sort of datum for a phylogeny of knowledge. The value is also to capture the visions and the styles of the scientists who preceded us.

Linda Price Thomson
The Common but Less Frequent Loon and Other Essays
Chapter 10 (p. 69)

In 1995 I needed a book of statistical quotations. When I went to the library I found that there wasn't such a book and so decided to compile my own. That book of statistical quotations has now grown into a series of six science quotation books. *Scientifically Speaking: A Dictionary of Quotations,* the sixth book of the series, is the largest compilation of published science quotations available to readers. It is a book that is designed to be entertaining but not authoritive; informative but not instructive. The purpose of the book is to present quotations so that the reader can gain an idea as to the depth and breadth of the subject of science and, also, to help the reader 'capture the visions and the styles of the scientists' of both the past and of the present. The bibliography is intended for those individuals who may wish to search for more details about the quotations listed.

With so many well-written books of quotations on the market is another book of quotations really needed? We and our publisher agreed that there was a need since the standard dictionaries of quotations, for whatever cause, are sorely weak in providing entries devoted to quotations on science. *Scientifically Speaking* fills that need.

The understanding of the history, the accomplishments and failures, and the meaning of science requires a knowledge of what has been said by the authoritative and the not so authoritative philosophers, novelists, playwrights, poets, scientists and laymen about science. Because of the multidisciplinary interrelationships that exist it is virtually impossible for an individual to keep abreast of the literature outside of their own particular specialization. With this in mind, *Scientifically Speaking* contains the words and wisdom of several hundred individuals—scientists, writers, philosophers, poets and academicians. The book assumes a particularly important role as a guide to what has been said from the past through to the present about science.

Perhaps the reader may ask, of what consequence is it whether the author's exact language is preserved or not, provided we have his thought? The answer is, that inaccurate quotation is a sin against truth. It may appear in any particular instance to be a trifle, but perfection consists in small things, and perfection is no trifle.

Robert W Shaunon
The Canadian Magazine
Misquotation
October 1898

Scientifically Speaking was designed as an aid for the general reader who has an interest in science topics as well as for the experienced scientist. The general reader with no knowledge of science who reads *Scientifically Speaking* can form a pretty accurate picture of what science is. Students can use the book to increase their understanding of the complexity and richness that exists within the scientific disciplines. Finally, the experienced scientist will find *Scientifically Speaking* useful as a source of quotes for use in the classroom, in papers and in presentations. We have striven to compile the book so that any reader can easily and quickly access the wit and wisdom that exists and a quick glance through the table of contents will show the variety of topics discussed.

A book of quotations, even as restricted in scope as *Scientifically Speaking* is, can never be complete. Many quotations worthy of entry have, no doubt, been omitted because we did not know of them. However, we have tried to make it fairly comprehensive and have searched far and wide for the material. If you are aware of any quotes that should be included please send them in for the second edition.

Quite a few of the quotations have been used frequently and will be recognized while others have probably not been used before. All of the quotations included in *Scientifically Speaking* were compiled with the hope that they will be found useful. The authority for each quotation has been given with the fullest possible information that we could find so as to help you pinpoint the quotation in its appropriate context or discover more quotations in the original source. When the original source could not be located we indicated where we found the quote. Sometimes, however, we only had the quote and not the source. When this happened we listed the source as unknown and included the quotation anyway so that it would not become lost in time.

How to Use This Book

1. A quotation for a given subject may be found by looking for that subject in the alphabetical arrangement of the book itself. This arrangement will be approved, we believe, by the reader in making it easier to locate a quotation. To illustrate, if a quotation on 'Abstraction' is wanted, you will find nine quotations listed under the heading ABSTRACTION. The arrangement of quotations in this book under each subject heading constitutes a collective composition that incorporates the sayings of a range of people.
2. To find all the quotations pertaining to a subject and the individuals quoted use the SUBJECT BY AUTHOR INDEX. This index will help guide you to the specific statement that is sought. A brief extract of each quotation is included in this index.
3. It will be admitted that at times there are obvious conveniences in an index under author's names. If you recall the name appearing in the attribution or if you wish to read all of an individual author's contributions that are included in this book then you will want to use the AUTHOR BY SUBJECT INDEX. Here the authors are listed alphabetically along with their quotations. The birth and death dates are provided for the authors whenever we could determine them.

Thanks

It is never superfluous to say thanks where thanks are due. Firstly, we want to thank Jim Revill and Al Troyano, of Institute of Physics Publishing, who have assisted us so very much with our books. Next, we thank the following libraries for allowing us to use their resources: The Jesse H. Jones Library and the Moody Memorial Library, Baylor University; the main library of the University of Mary-Hardin Baylor; the main library of the Central Texas College; the Perry-Castañeda Library, the Undergraduate Library, the Engineering Library, the Law Library, the Physics-Math-Astronomy Library, and the Humanities Research Center, all of the University of Texas at Austin. Again, we wish to thank Joe Gonzalez, Matt Pomeroy, Debbie Frank, JoAnna Cook, Chris Braun, Ken McFarland, Craig McDonald, Kathryn Kenefick, Brian Camp, Robert Clontz, and Gabriel Alvarado of the Perry-Castañeda Library for putting up with us when we were checking out the hundreds of books. Finally, we wish to thank our children Maritza, Maurice, and Marilynn for their assistance in finding the books we needed when we were at the libraries.

A great amount of work goes into the preparation of any book. When the book is finished there is then time for the editors and authors to enjoy what they have written. It is hoped that this book will stimulate your imagination and interests in matters about science and the scientific disciplines, and this hope has been eloquently expressed by Helen Hill (quoted in Llewellyn Nathaniel Edwards *A Record of History and Evolution of Early American Bridges*, p. xii):

> If what we have within our book
> Can to the reader pleasure lend,
> We have accomplished what we wished,
> Our means have gained our end.

Carl Gaither
Alma Cavazos-Gaither
January 2000

ABSTRACTION

Bacon, Francis

Although the roads to human power and to human knowledge lie close together, and are nearly the same, nevertheless on account of the pernicious and inveterate habit of dwelling on abstractions, it is safer to begin and raise the sciences from those foundations which have relation to practice, and to let the active part itself be as the seal which prints and determines the contemplative counterpart.

Novum Organum
Aphorism IV

Dingle, Herbert

Abstraction is the detection of a common quality in the characteristics of a number of diverse observations: it is the method supremely exemplified in the work of Newton and Einstein . . .

A hypothesis serves the same purpose, but in a different way. It relates apparently diverse experiences, not by directly detecting a common quality in the experiences themselves, but by inventing a fictitious substance or process or idea, in terms of which the experience can be expressed. A hypothesis, in brief, correlates observations by adding something to them, while abstraction achieves the same end by subtracting something.

Science and Human Experience (pp. 22–3)

Haber, Fritz

The field of scientific abstraction encompasses independent kingdoms of ideas and of experiments and within these, rulers whose fame outlasts the centuries.

In Richard Willstätter
From My Life
Chapter 8 (p. 174)

Huxley, Aldous

Knowledge is power and, by a seeming paradox, it is through their knowledge of what happens in this unexperienced world of abstractions and inferences that scientists and technologists have acquired their enormous and growing power to control, direct and modify the world of manifold appearances in which human beings are privileged and condemned to live.

Literature and Science
Chapter 3 (p. 9)

Joubert, Joseph
How many people become abstract in order to appear profound! Most abstract terms are shadows that conceal a void.

Pensées
XI (p. 88)

More, Louis Trenchard
The goal of science is mathematics, and while mathematics may be said to be the only true science since it has the only true scientific method, mathematics is not a science because it deals with abstractions and ignores concrete phenomena.

The Limitations of Science
Chapter V (p. 151)

Russell, Bertrand
The power of using abstractions is the essence of intellect, and with every increase in abstraction the intellectual triumphs of science are enhanced.

The Scientific Outlook
Chapter III (p. 87)

Sullivan, J.W.N.
Science, indeed, tells us a very great deal less about the universe than we have been accustomed to suppose, and there is no reason to believe that all we can ever know must be couched in terms of its thin and largely arbitrary abstractions.

Beethoven, His Spiritual Development
Art and Reality (pp. 21–2)

Whitehead, Alfred North
Matter-of-fact is an abstraction, arrived at by confining thought to purely formal relations which then masquerade as the final reality. This is why science, in its perfection, relapses into the study of differential equations. The concrete world has slipped through the meshes of the scientific net.

Modes of Thought (p. 25)

. . . the utmost abstractions are the true weapons with which to control our thought of concrete fact.

Science and the Modern World
Chapter II (p. 32)

How many people become abstract in order to appear profound!
Joseph Joubert – (See p. 2)

ADMINISTRATION OF SCIENCE

Mellanby, Kenneth

. . . the corridors of power have a strong attraction for even the most devoted investigator, and these corridors seldom lead back to the laboratory.

New Scientist
Disorganisation of Scientific Research (p. 436)
Volume 59, Number 86023, August 1973

Wiener, Norbert

There are many administrators of science and a large component of the general population who believe that mass attacks can do anything, and even that ideas are obsolete. Behind this drive to the mass attack there are a number of strong psychological motives. Neither the public nor the big administrator has too good an understanding of the inner continuity of science, but they have both seen its world-shaking consequences, and they are afraid of it. Both of them wish to decerebrate the scientist, as the Byzantine State emasculated its civil servants. Moreover the great administrator who is not sure of his own intellectual level can aggrandise himself only by cutting his scientific employees down to size.

I Am A Mathematician
Epilogue (p. 363)

AESTHETICS

Bragg, W.L.

When one has sought long for the clue to a secret of nature, and is rewarded by grasping some part of the answer, it comes as a blinding flash of revelation: it comes as something new, more simple and at the same time more aesthetically satisfying than anything one would have created in one's own mind. This conviction is of something revealed, and not something imagined.

In C.A. Coulson
Science and Christian Belief
Christian Belief (p. 99)

Kline, Morris

Much research for new proofs of theorems already correctly established is undertaken simply because the existing proofs have no aesthetic appeal. There are mathematical demonstrations that are merely convincing; to use a phrase of the famous mathematical physicist, Lord Rayleigh, they 'command assent.' There are other proofs 'which woo and charm the intellect. They evoke delight and an overpowering desire to say, Amen, Amen.' An elegantly executed proof is a poem in all but the form in which it is written.

Mathematics in Western Culture (p. 470)

Kuhn, Thomas S.

The importance of aesthetic considerations can sometimes be decisive. Though they often attract only a few scientists to a new theory, it is upon those few that its ultimate triumph may depend. If they had not quickly taken it up for highly individual reasons, the new candidate for paradigm might never have been sufficiently developed to attract the allegiance of the scientific community as a whole.

The Structure of Scientific Revolutions (p. 156)

Penrose, Roger

Aesthetic qualities are important in science, and necessary, I think, for great science.

Scientific American
In John Horgan
Quantum Consciousness (p. 32)
Volume 261, Number 5, November 1989

Poincaré, Henri

And it is because simplicity, because grandeur, is beautiful, that we preferably seek simple facts, sublime facts, that we delight now to follow the majestic course of the stars, not to examine with the microscope that prodigious littleness which is also a grandeur, now to seek in geologic time the traces of a past which attracts because it is far away.

The Foundations of Science
Science and Method
The Choice of Facts (p. 23)

Sullivan, J.W.N.

Since the primary object of the scientific theory is to express the harmonies which are found to exist in nature, we see at once that these theories must have an aesthetic value. The measure of the success of a scientific theory is, in fact, a measure of its aesthetic value, since it is a measure of the extent to which it has introduced harmony in what was before chaos.

The Athenaeum
The Justification of the Scientific Method (p. 275)
Number 4644, May 2, 1919

Thomson, Sir George

One can always make a theory, many theories, to account for known facts, occasionally even to predict new ones. The test is aesthetic.

The Inspiration of Science
The Scientific Method (p. 17)

AGE OF SCIENCE

Compton, Karl Taylor

We live in an age of science. I do not say "an age of technology" for every age has been an age of technology. We recognize this when we describe past civilizations as the Stone Age, the Bronze Age, and the Age of Steam or of Steel, thus implicitly admitting that the stage of civilization is determined by the tools at man's disposal—in other words, by his technology . . . Science, unlike invention and technical skill, is a relatively modern concept.

A Scientist Speaks (p. 1)

Kronenberger, L.

Nominally a great age of scientific inquiry, ours has actually become an age of superstition about the infallibility of science; of almost mystical faith in its nonmystical methods; above all . . . of external verities; of traffic-cop morality and rabbit-test truth.

Company Manners
Chapter 4 (p. 94)

Russell, Bertrand

So far I have been speaking of *theoretical* science, which is an attempt to *understand* the world. *Practical* science, which is an attempt to *change* the world, has been important from the first, and has continually increased in importance, until it has almost ousted theoretical science from men's thoughts . . . The triumph of science has been mainly due to its practical utility, and there has been an attempt to divorce this aspect from that of theory, thus making science more and more a technique, and less and less a doctrine as to the nature of the world. The penetration of this point of view to philosophers is very recent.

History of Western Philosophy
Book Three
Chapter I (p. 480)

To say that we live in an age of science is a common place, but like most common places, it is only partially true. From the point of view of our predecessors, if they could view our society, we should, no doubt, appear to be very scientific, but from the point of view of our successors, it is probable that the exactly opposite would seem to be the case.

The Scientific Outlook
Introduction (p. 9)

. . . the corridors of power have a strong attraction for even the most devoted investigator, and these corridors seldom lead back to the laboratory.
Kenneth Mellanby – (See p. 4)

ANALOGY

Campbell, Norman R.

. . . analogies are not "aids" to the establishment of theories; they are an utterly essential part of theories, without which theories would be completely valueless and unworthy of the name. It is often suggested that the analogy leads to the formulation of the theory, but once the theory is formulated the analogy has served its purpose and may be removed and forgotten. Such a suggestion is absolutely false and perniciously misleading.

Physics, The Elements
Chapter VI (p. 129)

To regard analogy as an aid to the invention of theories is as absurd as to regard melody as an aid to the composition of sonatas.

Physics, The Elements
Chapter VI (p. 130)

Cohen, Morris R.

. . . the number of available analogies is a determining factor in the growth and progress of science.

The Meaning of Human History
Chapter 8 (p. 249)

Hesse, Mary B.

. . . one of the main functions of an analogy or model is to suggest extensions of the theory by considering extensions of the analogy, since more is known about the analogy than is known about the subject matter of the theory itself . . . A collection of observable concepts in a purely formal hypothesis suggesting no analogy with anything would consequently not suggest either any directions for its own development.

British Journal for the Philosophy of Science
Operational Definition and Analogy in Physical Theories (p. 291)
Volume II, Number 8, February 1952

Latham, Peter

It is safest and best to fill up the gap of our knowledge from analogy.

In William Bean
Aphorisms from Latham (p. 37)

Thoreau, Henry David

All perception of truth is the detection of an analogy . . .

The Journal of Henry D. Thoreau
Volume II
September 5, 1851 (p. 463)

Westbrock, Peter

We have known since the days of Kant that scientific arguments must never be founded on analogies, but the authors are dead serious about these poetic digressions.

The Times Higher Education Supplement
The Oceans Inside Us
November 3, 1995

ANSWER

McKuen, Rod

Think of all the men who never knew the answers
think of all those who never even cared.
Still there are some who ask why
who want to know, who dare to try.

Listen to the Warm Poems
Here He Comes Again

Pasteur, Louis

Science proceeds by successive answers to questions more and more subtle, coming nearer and nearer to the very essence of phenomena.

Études sur la Bière

ANTI-SCIENCE

Anderson, Maxwell

. . . if you live you have to be going somewhere. You have to choose a direction. And science is completely impartial. It doesn't give a damn which way you go. It can invent the atom bomb but can't tell whether to use it or not. Science is like—well, it's like a flashlight in a totally dark room measuring two billion light-years across—and with walls that shift away from you as you go towards them. The flash can show you where your feet are on the floor; it can show you the furniture or the people close by; but as for which direction you should take in that endless room it can tell you nothing.

Joan of Lorraine
Act II, Rehearsal Preface (p. 83)

Asimov, Isaac

A public that does not understand how science works can, all too easily, fall prey to those ignoramuses . . . who make fun of what they do not understand, or to the sloganeers who proclaim scientists to be the mercenary warriors of today, and the tools of the military. The difference . . . between . . . understanding and not understanding . . . is also the difference between respect and admiration on the one side, and hate and fear on the other.

In Lewis Wolpert
The Unnatural Nature of Science
Introduction (p. ix)

France, Anatole

I hate science . . . for having loved it too much, after the manner of voluptuaries who reproach women with not having come up to the dream they formed of them.

The Authorized English Translation of the Novels and Short Stories of Anatole France
Volume II
The Opinions of Jérôme Coignard
Chapter 9 (p. 113)

Gissing, George
I hate and fear science because of my conviction that for long to come if not for ever, it will be the remorseless enemy of mankind. I see it destroying all simplicity and gentleness of life, all the beauty of the world; I see it restoring barbarism under a mask of civilization; I see it darkening men's minds and hardening their hearts; I see it bringing a time of vast conflicts which will pale into insignificance 'the thousand wars of old' and, as likely as not, will wheel all the laborious advances of mankind in blood-drenched chaos.

In Morris Goran
Science and Anti-Science
Chapter 3 (p. 23)

Green, Celia
The object of modern science is to make all aspects of reality equally boring, so that no one will be tempted to think about them.

The Decline and Fall of Science
Aphorisms (p. 2)

Hardy, Thomas
Well: what we gain by science is, after all, sadness, as the Preacher saith. The more we know of the laws & nature of the Universe the more ghastly a business we perceive it all to be—& the non-necessity of it.

Collected Letters
Volume 3
Letter to Edward Clodd
February 27, 1902 (p. 5)

APPLIED SCIENCE

Bacon, Francis

Even when men build any science and theory upon experiment, yet they almost always turn with premature and hasty zeal to practice, not merely on account of the advantage and benefit to be derived from it, but in order to seize upon some security in a new undertaking of their not employing the remainder of their labor unprofitably, and by making themselves conspicuous, to acquire a praetor name for their pursuit.

Novum Organum
First Book, 70

Compton, Karl Taylor

Applied science is not an end in itself, but it is the most powerful means ever discovered for supplying the opportunity to secure the finest things of life.

A Scientist Speaks (p. 9)

Einstein, Albert

It is not enough that you should understand about applied science in order that your work may increase man's blessings. Concern for the man himself and his fate must always form the chief interest of all technical endeavors; concern for the great unsolved problems of the organization of labor and the distribution of our mind shall be a blessing and not a curse to mankind.

Never forget this in the midst of your diagrams and equations.

The NY Times
Einstein Seeks Lack in Applying Science (p. 6)
February 17, 1931

Why does this magnificent applied science which saves work and makes life easier bring us so little happiness? The simple answer runs: Because we have not yet learned to make sensible use of it.

The NY Times
Einstein Seeks Lack in Applying Science (p. 6)
February 17, 1931

Huxley, Aldous
Applied Science is a conjurer, whose bottomless hat yields impartially the softest of Angora rabbits and the most petrifying of Medusas.

Tomorrow and Tomorrow and Tomorrow
The Desert (p. 85)

Huxley, Thomas
I often wish that this phrase "applied science", had never been invented. For it suggests that there is a sort of scientific knowledge of direct practical use, which can be studied apart from another sort of scientific knowledge, which is of no practical utility, and which is termed "pure science". But there is no more complete fallacy than this.

Collected Essays
Volume III
Science and Education
Science and Culture (p. 137)

Pasteur, Louis
There does not exist a category of science to which one can give the name applied science. There are science and the applications of science, bound together as the fruit of the tree which bears it.

Revue Scientifique
Pourquoi la France n'a pas trouvé hommes supérieurs
au montant du péril (1871)

Porter, Sir George
To feed applied science by starving basic science is like economising on the foundations of a building so that it may be built higher.

New Scientist
Lest the Edifice of Science Crumble (p. 16)
Volume 11, Number 1524, September 1986

Thomas, Lewis
[Surprise] is the element that distinguishes applied science from basic . . . When you are organized to apply knowledge, set up targets, produce a usable product, you require a high degree of certainty from the outset. All the facts on which you base protocols must be reasonably hard facts with unambiguous meaning. The challenge is to plan the work and organize the workers so that it will come out precisely as predicted. For this, you need centralized authority, elaborately detailed time schedules, and some sort of reward system based on speed and perfection. But most of all you need the intelligible basic facts to begin with, and these must come from basic research. There is no other source.

In basic research, everything is just the opposite. What you need at the outset is a high degree of uncertainty; otherwise it isn't likely to be an important problem. You start with an incomplete roster of facts, characterized by their ambiguity; often the problem consists of discovering the connections between unrelated pieces of information. You must plan experiments on the basis of probability, or even bare possibility, rather than certainty. If an experiment turns out precisely as predicted this can be very nice, but it is only a great event if at the same time it is a surprise. You can measure the quality of the work by the intensity of astonishment.

The Lives of a Cell
The Planning of Science (pp. 118–9)

Wheeler, Edgar C.

. . . they cannot remain indefinitely in any field of pure research. For every time they come upon a new bit of knowledge, almost instantly they discover some practical application. Thus the dividing line between pure science and applied science becomes thin.

The World's Work
Makers of Lightning (p. 271)
January 1927

AXIOMS

Chargaff, Erwin

. . . nowadays our sciences, quick and fickle, wear out dogmas in 10 years, and axioms take only a little longer.

Perspectives in Biology and Medicine
Bitter Fruits from the Tree of Knowledge
Section VIII (p. 496)
Volume 16, Number 4, Summer 1973

Raju, P.T.

We are driven to conclude that science, like mathematics, is a system of axioms, assumptions, and deductions; it may start from being, but later leaves it to itself, and ends in the formation of a hypothetical reality that has nothing to do with existence; or it is the discovery of an ideal being which is of course, present in what we call actuality, and renders it an existence for us only by being present in it.

Idealistic Thought of India (p. 84)

BEAUTY

Curie, Marie
I am among those who think that science has great beauty . . . A scientist in his laboratory is not only a technician but also a child placed in front of natural phenomena which impresses him like a fairy tale.

In Eve Curie
Madame Curie
Full Bloom (p. 341)

Dretske, Fred I.
Beauty is in the eye of the beholder, and information is in the head of the receiver.

Knowledge & the Flow of Information
Preface (p. vii)

Kragh, Helge
The main problem is that beauty is essentially subjective and hence cannot serve as a commonly defined tool for guiding or evaluating science. It is, to say the least, difficult to justify aesthetic judgment by rational arguments . . . The sense of aesthetic standards is part of the socialization that scientists acquire; but scientists, as well as scientific communities, may have widely different ideas of how to judge the aesthetic merit of a particular theory. No wonder that eminent physicists do not agree on which theories are beautiful and which are ugly.

Dirac: A Scientific Biography
Chapter 14 (pp. 287–8)

Oppenheimer, J. Robert
The profession I'm part of has as its whole purpose, the rendering of the physical world understandable and beautiful. Without this you have only tables and statistics.

Look
With Oppenheimer on an Autumn Day (p. 63)
Volume 30, Number 26, December 27, 1966

Science is not everything. But science is very beautiful.

Look
With Oppenheimer on an Autumn Day (p. 63)
Volume 30, Number 26, December 27, 1966

Weil, Simone
The true definition of science is this: the study of the beauty of the world.

The Need for Roots
Part Three (p. 261)

There exists today a universal language . . .
it is about time that Broken English be regarded
as a language in its own right.
Hendrik B.G. Casimir – (See p. 26)

BOOKS

Heaviside, Oliver

The following story is true. There was a little boy, and his father said, "Do try to be like other people. Don't frown." And he tried and tried, but could not. So his father beat him with a strap; and then he was eaten up by lions.

Reader, if young, take warning by his sad life and death. For though it may be an honour to be different from other people, if Carlyle's dictum about the 30 million be still true, yet other people do not like it. So, if you are different, you had better hide it, and pretend to be solemn and wooden-headed. Until you make your fortune. For most wooden-headed people worship money; and, really, I do not see what else they can do. In particular, if you are going to write a book, remember the wooden-headed. So be rigorous; that will cover a multitude of sins. And do not frown.

Electromagnetic Theory
Volume III (p. 1)

Huxley, Thomas

. . . books are the money of Literature, but only the counters of Science.

Collected Essays
Volume III
Science and Education
Universities: Actual and Ideal (p. 213)

Richards, I.A.

A book is a machine to think with . . .

Principles of Literary Criticism
Preface (p. 1)

Slosson, E.E.

One obstacle in the way of spreading science, that is of inculcating the scientific habit of mind, is that people have learned to read too well.

Books may become an impediment to learning. Our students are taught how to learn to read but not always how to read to learn.

Digest of the Proceedings of the Second Annual Meeting of the American Association for Adult Education
Adult Education in Science (p. 53)
1927

Twain, Mark

There are three infallible ways of pleasing an author, and the three form a rising scale of compliment: 1—to tell him you have read one of his books; 2—to tell him you have read all of his books; 3—to ask him to let you read the manuscript of his forthcoming book. No. 1 admits you to his respect; No. 2 admits you to his admiration; No. 3 carries you clear into his heart.

The Tragedy of Pudd'nhead Wilson
Chapter XI

Wittgenstein, Ludwig

The popular scientific books by our scientists aren't the outcome of hard work, but are written when they are resting on their laurels.

Culture and Value (p. 42e)

CHAOS

Adams, Henry

Briefly chaos is all that science can logically assert of the supersensuous.

The Education of Henry Adams
The Grammar of Science (p. 451)

. . . Chaos was the law of nature; Order was the dream of man.

The Education of Henry Adams
The Grammar of Science (p. 451)

Blackie, John Stuart

Chaos, Chaos, infinite wonder!
Wheeling and reeling on wavering wings; . . .

Musa Burschicosa
A Song of Geology

Bronk, Detlev W.

Science, like art, music and poetry, tries to reduce chaos to the clarity and order of pure beauty.

Journal of the American Medical Association
In Max Levin
Our Debt to Hughlings Jackson (p. 996)
Volume 191, Number 12, March 22, 1965

George, William H.

One man's explanation may be another man's chaos.

The Scientist in Action
Personal Basis (p. 331)

Kant, Immanuel

. . . God has put a secret art into the forces of nature so as to enable it to fashion itself out of chaos into a perfect world system . . .

Universal Natural History and Theory of the Heavens
Preface (p. 27)

COMMON SENSE

Ackoff, Russell Lincoln
. . . common sense . . . has the very curious property of being more correct retrospectively than prospectively. It seems to me that one of the principal criteria to be applied to successful science is that its results are almost always obvious retrospectively; unfortunately they seldom are prospectively. Common sense provides a kind of ultimate validation after science has completed its work; it seldom anticipates what science is going to discover.

In Anthony de Reuck, Maurice Goldsmith and Julie Knight (Editors)
Decision Making in National Science Policy
Operational Research and National Science Policy
Discussion (p. 96)

Davy, Sir Humphrey
In the progress of an art, from its rudest to its more perfect state, the whole process depends upon experiments. Science is in fact nothing more than the refinement of common sense making use of facts already known to acquire new facts.

Consolations in Travel
Dialogue V
The Chemical Philosopher (p. 163)

Dewey, John
. . . unless the materials involved can be traced back to the material of common sense concern there is nothing whatever for scientific concern to be concerned with.

The Journal of Philosophy
Common Sense and Science (p. 206)
Volume XLV, Number 8, April 18, 1948

Holmes, Oliver Wendell

Science is a first-rate piece of furniture for a man's upper chamber, if he has common sense on the ground-floor.

The Poet at the Breakfast-Table (p. 141)

Huxley, Thomas

Science is, I believe, nothing but *trained and organized common sense*, differing from the latter only as a veteran may differ from a raw recruit: and its methods differ from those of common sense only as far as the guardsman's cut and thrust differ from the manner in which a savage wields a club.

Collected Essays
Volume III
Science and Education
On The Educational Value of the Natural History of Science (p. 45)

Lehrer, Keith

The overthrow of accepted opinion and the dictates of common sense are often essential to epistemic advance. Moreover, an epistemic adventurer may arrive at beliefs that are not only new and revelatory, but also better justified than those more comfortably held by others. The principle of the conservation of accepted opinion is a roadblock to inquiry, and, consequently, it must be removed.

Knowledge
Chapter 7 (p. 184)

Luria, S.E.

Significant advances in science often have a peculiar quality: they contradict obvious, commonsense opinions.

A Slot Machine, A Broken Test Tube: An Autobiography
The Science Path: IV
Looking Back (p. 116)

Oppenheimer, J. Robert

. . . distrust all the philosophers who claim that by examining science they come to the results in contradiction with common sense. Science is based on common sense; it cannot contradict it.

Foundations for World Order
The Scientific Foundations for World Order (p. 51)

Common sense is not wrong in the view that it is meaningful, appropriate and necessary to talk about the large objects of our daily experience . . . Common sense is wrong only if it insists that what is familiar must reappear in what is unfamiliar.

Science and the Common Understanding
Uncommon Sense (pp. 74–5)

Popper, Karl R.
All science and all philosophy are enlightened common sense.

Objective Knowledge (p. 34)

Thomson, J.A.
. . . one of the most marked characteristics of science is its critical quality, which is just what common-sense lacks. By common-sense is usually meant either the consensus of public opinion, of unsystematic everyday thinking, the untrustworthiness of which is notorious, or the verdict of uncritical sensory experience, which has so often proved fallacious. It was "common-sense" that kept the planets circling around the earth; it was "common-sense" that refused to accept Harvey's demonstration of the circulation of the blood.

Introduction to Science
Chapter II (p. 39)

Titchener, E.B.
Common sense is the very antipodes of science.

Systematic Psychology
Science and Logic (p. 48)

Whitehead, Alfred North
Now in creative thought common sense is a bad master. Its sole criterion for judgment is that the new ideas shall look like the old ones. In other words it can only act by suppressing originality.

An Introduction to Mathematics
Chapter 11 (p. 116)

Wolpert, Lewis
. . . one of the strongest arguments for the distance between common sense and science is that the whole of science is totally irrelevant to people's day-to-day lives.

The Unnatural Nature of Science
Chapter I (p. 16)

COMMUNICATION IN SCIENCE

Casimir, Hendrik B.G.

There exists today a universal language that is spoken and understood almost everywhere: it is Broken English. I am not referring to Pidgin-English—a highly formalized and restricted branch of B.E.—but to the much more general language that is used by the waiters in Hawaii, prostitutes in Paris and ambassadors in Washington, by businessmen from Buenos Aires, by scientists at international meetings and by dirty-postcard peddlers in Greece—in short, by honorable people like myself all over the world . . . The number of speakers of Broken English is so overwhelming and there are so many for whom B.E. is almost the only way of expressing themselves—at least in certain spheres of activity—that it is about time that Broken English be regarded as a language in its own right.

Haphazard Reality
Chapter 4
Broken English (p. 122)

Chargaff, Erwin

. . . there is no real popularization possible, only vulgarization that in most instances distorts the discoveries beyond recognition.

Perspectives in Biology and Medicine
Bitter Fruits from the Tree of Knowledge
Section III (p. 491)
Volume 16, Number 4, Summer 1973

Compton, Karl Taylor

The whole history of scientific progress illustrates the importance of free communication of ideas, of co-operative work at all levels, of adequate support and facilities, and above all, of high grade research workers and top-notch leadership.

A Scientist Speaks (p. 11)

Dornan, Christopher

Science is seen as an avenue of access to assured findings, and scientists—in the dissemination of these findings—as the initial sources. The members of the laity are understood purely as recipients of this information. Journalists and public relations personnel are viewed as intermediaries through which the scientific findings filter. The task for science communication is to transmit as much information as possible with maximum fidelity.

Critical Studies in Mass Communication
Some Problems of Conceptualizing the Issues of
"Science and the Media" (p. 51)
Volume 7, Number 1, March 1990

Gibbs, Willard

Science is, above all, communication.

Attributed
In H.N. Parton
Science is Human
Science and the Liberal Arts (p. 11)

Ingle, Dwight J.

Science cannot be equated to measurement, although many contemporary scientists behave as though it can. For example, the editorial policies of many scientific journals support the publication of data and exclude the communication of ideas.

Principles of Research in Biology and Medicine
Chapter 1 (p. 3)

Maslow, A.H.

I do not recall seeing in the literature with which I am familiar, any paper that criticized another paper for being unimportant, trivial or inconsequential.

Motivation and Personality
Chapter 2 (p. 14)

Michelson, A.A.

Science, when it has to communicate the results of its labor, is under the disadvantage that its language is but little understood. Hence it is that circumlocution is inevitable and repetitions are difficult to avoid.

Light Waves and Their Uses
Lecture I (p. 1)

Moore, John A.

. . . recall some of the lectures you may have heard recently. Did you always know why the research had been done? Was it clear what problem was being illuminated by the data presented?

American Zoologist
Science as a Way of Knowing (p. 471)
Volume 24, Number 2, 1984

Moravcsik, M.J.

New theories, when first proposed, may appear on the first page of the *New York Times*, but their demise, a few years later, never makes even page 68.

Research Policy
Volume 17, 1988 (p. 293)

Parton, H.N.

Scientists have the duty to communicate, firstly with each other, that is with those who are interested in the same or allied problems, and secondly with laymen: by layman I mean anyone not familiar with their special science, for specialization has raised the level of scientific achievement so much, that chemists, for example, are usually laymen in say, biology; we may hope, intelligent laymen.

Science is Human
Science and the Liberal Arts (p. 12)

Aldous Huxley, in a lecture on his grandfather, said that all communication is literature, and even in scientific writing there is wide room for the exercise of art.

Science is Human
Science and the Liberal Arts (p. 14)

Roe, Anne

Nothing in science has any value to society if it is not communicated . . .

The Making of a Scientist
Chapter I (p. 17)

Seifriz, William

Our scientific congresses are a hodgepodge of trivia. The conversation is that of men on the defensive.

Science
A New University (p. 89)
Volume 120, Number 3107, July 16, 1954

Wang, H.

. . . face-to-face discussion, complete with gestures, not only transmit ideas much more rapidly between specialists but also generally gets more directly to the fuller implications of ideas.

Perspectives in Biology and Medicine
The Formal and the Intuitive in the Biological Sciences (p. 528)
Volume 27, Number 4, 1984

Ziman, John

Although the best and most famous scientific discoveries seem to open whole new windows of the mind, a typical scientific paper has never pretended to be more than another little piece in a larger jigsaw—not significant in itself but as an element in a grander scheme. This technique, of soliciting many modest contributions to the vast store of human knowledge, has been the secret of Western science since the seventeenth century, for it achieves a corporate, collective power that is far greater than any one individual can exert. Primary scientific papers are not meant to be final statements of indisputable truths; each is merely a tiny tentative step forward, through the jungle of ignorance.

Nature
Information, Communication, Knowledge (p. 324)
Volume 224, Number 5217, October 25, 1969

It is not enough to observe, experiment, theorize, calculate and communicate; we must also argue, criticize, debate, expound, summarize, and otherwise transform the information that we have obtained individually into reliable, well established, public knowledge.

Nature
Information, Communication, Knowledge (p. 324)
Volume 224, Number 5217, October 25, 1969

The cliché of scientific prose betrays itself 'Hence *we* arrive at the conclusion that . . .' The audience to which scientific publications are addressed is not passive; by its cheering or booing, its bouquets or brickbats, it actively controls the substance of the communications that it receives.

Public Knowledge
Chapter 1 (p. 9)

CONCEPTS

Conant, James Bryant

. . . a useful concept may be a barrier to the acceptance of a better one if long-intrenched in the minds of scientists.

On Understanding Science
Chapter III (p. 74)

Egler, Frank E.

A concept is nothing more than an idea, a mental creation, which makes comprehensible a certain group of facts.

The Way of Science
Science Concepts (p. 21)

Concepts are games we play with our heads; methods are games we play with our hands, which at times are so handy they can be played without a head.

The Way of Science
The Nature of Science (p. 1)

Riemann, Bernhard

Science is the attempt to comprehend nature by means of concepts.

In C.J. Keyser
The Hibbert Journal
Volume 3, 1904–1905 (pp. 312–3)

Tansley, A.G.

We must never conceal from ourselves that our concepts are creations of the human mind which we impose on the facts of nature, that they are derived from incomplete knowledge, and therefore will never *exactly* fit the facts, and will require constant revision as knowledge increases.

Journal of Ecology
Classification of Vegetation and the Concept of Development (p. 120)
Volume 8, 1920

DATA

Fox, Russell
Gorbuny, Max
Hooke, Robert

It has been said that data collection is like garbage collection: *before* you collect it you should have in mind what you are going to do with it.

The Science of Science
Chapter 6 (p. 51)

Inose, Hiroshi
Pierce, J.R.

Where is the information we have lost in data?

Information Technology and Civilization
Chapter 7-4 (p. 210)

Katsaros, Kristina

Sometimes there are heated arguments at meetings about how to interpret data. When you have very few facts, fully interpreting them can give rise to three or four interpretations—within the error bars, the uncertainties in the measurements. You get people adhering to one or the other interpretation for a while, and that's not based on fact because there are not enough facts. Eventually more facts are gathered and it becomes clear what the answer is, and everybody agrees. In the end you have a new result. That's the wonderful thing about science, that you can only find in science. There is a point when there is no doubt anymore. There is usually a lot of emotional stress before you get rid of some former idea. There may be a few crackpots who fight it, but if the evidence is good, eventually all accept it. I think that's wonderful. One of the best things about science is that there are some objective answers.

In Linda Jean Shepherd
Lifting the Veil
Receptivity (pp. 99–100)

DEFINITION

Arnold, Thurman

Definition is ordinarily supposed to produce clarity in thinking. It is not generally recognized that the more we define our terms the less descriptive they become and the more difficulty we have in using them.

The Folklore of Capitalism
Chapter VII (p. 180)

Boutroux, Émile

There can be nothing clearer or more convenient for the purpose of setting one's ideas in order and for conducting an abstract discussion, than precise definitions and inviolable lines of demarcation.

Science & Religion in Contemporary Philosophy
Chapter I (p. 39)

Collingwood, R.G.

The beginner *has in his head a definition of the science*; a childish definition, perhaps, but still a definition; on the science's subject-matter he has *no definition at all*.

The New Leviathan
Part 1, Chapter 1, aphorism I.43

Einstein, Albert

Every physical concept must be given a definition such that one can in principle describe, in virtue of this definition, whether or not it applies in each particular case.

In Maurice Solovine
Lettres à Maurice Solovine (p. 20)

The strangest thing in all this medieval literature is the conviction that if there is a word there must also be a clear meaning behind it, and the only problem is to find out that meaning.

In James T. Cushing, C.F. Delaney and Gary M. Gutting
Science and Reality
Letter to Rabbi P.D. Bookstaber, August 24, 1951 (p. 108)

Hobbes, Thomas

. . . a man that seeketh precise truth, had need to remember what every name he uses stands for; and to place it accordingly; or else he will find himselfe entangled in words, as a bird in lime-twigs; the more he struggles, the more belimed.

Leviathan
Part I, Chapter IV (p. 56)

Huxley, Aldous

In the scientist verbal caution ranks among the highest of virtues. His words must have a one-to-one relationship with some specific class of data or sequence of ideas. By the rules of the scientific game he is forbidden to say more than one thing at a time, to attach more than one meaning to a given word, to stray outside the bounds of logical discourse, or to talk about his private experiences in relation to his work in the domains of public observation and public reasoning . . .

Literature and Science
Chapter 14 (p. 36)

Newton, Sir Isaac

My Design in this Book is not to explain the Properties of Light by Hypotheses, but to propose and prove by Reason and Experiments: In order to which I shall premise the following Definitions and Axioms.

Opticks
Book One, Part I (p. 379)

Ruse, Michael

It is simply not possible to give a neat definition—specifying necessary and sufficient characteristics—which separates all and only those things that have ever been called "science".

Science, Technology & Human Values
Response to the Commentary: *Pro Judice* (p. 72)
Volume 7, Number 41, Fall 1982

DISCOVERY

Bernard, Claude

A great discovery is a fact whose appearance in science gives rise to shining ideas, whose light dispels many obscurities and shows us new paths.

An Introduction to the Study of Experimental Medicine
Part I, Chapter II, Section ii (p. 34)

. . . a discovery is generally an unforeseen relation not included in theory, for otherwise it would be foreseen.

An Introduction to the Study of Experimental Medicine
Part I, Chapter II, Section iii (p. 38)

It has often been said that, to make discoveries, one must be ignorant. This opinion, mistaken in itself, nevertheless conceals a truth. It means that it is better to know nothing than to keep in mind fixed ideas based on theories whose confirmation we constantly seek, neglecting meanwhile everything that fails to agree with them.

An Introduction to the Study of Experimental Medicine
Part I, Chapter II, Section iii (p. 37)

Men who have excessive faith in their theories or ideas are not only ill prepared for making discoveries; they also make very poor observations.

An Introduction to the Study of Experimental Medicine
Part I, Chapter II, Section iii (p. 38)

Ardent desire for knowledge, in fact, is the one motive attracting and supporting investigators in their efforts; and just this knowledge, really grasped and yet always flying away before them, becomes at once their sole torment and sole happiness. Those who do not know the torment of the unknown cannot have the joy of discovery, which is certainly the liveliest that any man can feel.

An Introduction to the Study of Experimental Medicine
Part III, Chapter IV, Section iv (pp. 221–2)

Beveridge, W.I.B.
Often the original discovery, like the crude ore from the mine, is of little value until it has been refined and fully developed.

The Art of Scientific Investigation
Reason (p. 91)

Camras, Marvin
A scientific discovery . . . doesn't have to please anybody. It just has to be in accordance with nature, and it has to work.

In Kenneth A. Brown
Inventors at Work
Marvin Carmas (p. 75)

Chomsky, Noam
Discovery is the ability to be puzzled by simple things.

Chicago Tribune
In Ron Grossman
Strong Words
Asking the Questions
1:5, Section 5, January 1, 1993

Conant, James Bryant
. . . experimental discoveries must fit the time; facts may be at hand for years without their significance being realized; the total scientific situation must be favorable for the acceptance of new views.

On Understanding Science
Chapter III (p. 74)

Davy, Sir Humphrey
Imagination, as well as reason, is necessary to perfection in the philosophical mind. A rapidity of combination, a power of perceiving analogies, and of comparing them by facts, is the creative source of discovery.

In John Davy (Editor)
The Collected Works of Sir Humphrey Davy
Volume VIII
Parallels Between Art and Science (p. 308)

de Maistre, J.

Those who have made the most discoveries in science are those who knew Bacon least, while those who have read and pondered him, like Bacon himself, have not succeeded well.

In H. Selye
From Dream to Discovery
How to Think (p. 263)

Eden, Sir Anthony

Every succeeding scientific discovery makes greater nonsense of old-time conceptions of sovereignty.

Speech, House of Commons
November 22, 1945

Einstein, Albert

The use of the word 'Discovery' in itself is to be deprecated. For discovery is equivalent to becoming aware of a thing which is already formed; this links up with proof, which no longer bears the character of 'discovery' but, in the last instance, of the means that leads to discovery . . . Discovery is really not a creative act!

In Alexander Moszkowski
Conversations with Einstein
Chapter V (p. 95)

. . . the scientist finds his reward in what Henri Poincaré calls the joy of comprehension, and not in the possibilities of application to which any discovery of his may lead.

In Max Planck
Where is Science Going?
Epilogue (p. 211)

Epps, John

Let it be remembered that there are, in every discovery, two accidents; the accident of meeting with the fact connected with the discovery, and the accident of possessing an ingenious, and, in most cases, a great mind to take advantage of the fact.

Life of John Walker
Chapter XII (p. 295)

Fuller, R. Buckminster

Scientific discovery is invention, and *vice versa*.

Nine Chains to the Moon
Chapter 22 (p. 181)

Galton, Francis

I have been speculating about what makes a man a discoverer of undiscovered things . . . Many men who are very clever—much cleverer than the discoverers—never originate anything. As far as I can conjecture the art consists in habitually searching for the causes and meaning of everything which occurs. This implies sharp observation and requires as much knowledge as possible of the subject investigated.

In Karl Pearson
The Life, Letters and Labours of Francis Galton
Volume II
Letter to his son Horace
December 15, 1871 (p. 1)

Gay-Lussac, Joseph Louis

A discovery is the product of a previous discovery and in its turn it will give rise to a further discovery.

In Maurice Crosland
Gay-Lussac: Scientist and Bourgeois
Chapter 4 (p. 71)

Hamilton, Sir William Rowan

In physical sciences the discovery of new facts is open to any blockhead with patience and manual dexterity and acute senses.

In W.I.B. Beveridge
The Art of Scientific Investigation
Scientists (p. 144)

Harwit, Martin

The history of most efforts at discovery follows a common pattern, whether we consider the discovery of varieties of insects, the exploration of the oceans for continents and islands, or the search for oil reserves in the ground. There is an initial accelerating rise in the discovery rate as increasing numbers of explorers become attracted. New ideas and tools are brought to bear on the search, and the pace of discovery quickens. Soon, however, the number of discoveries remaining to be made dwindles, and the rate of discovery declines despite the high efficiency of the methods developed. The search is approaching an end. An occasional, previously overlooked feature can be found or a particular rare species encountered; but the rate of discovery begins to decline quickly and then diminishes to a trickle. Interest drops, researchers leave the field, and there is virtually no further activity.

Cosmic Discovery
Chapter I (pp. 42–3)

Huggins, William

From individual minds are born all great discoveries and revolutions of thought. New ideas may be in the air, and more or less present in many minds, but it is always an individual who at the last takes the creative step and enriches mankind with the living germ-thought of a new era of opinion.

Presidential address
Royal Society Anniversary Meeting
November 30, 1905
In William H. George
The Scientist in Action
Some Problems in Theorizing (p. 265)

Lovecraft, H.P.
de Castro, Adolphe
What a grown man worships is truth—knowledge—science—light—the rending of the veil and the pushing back of the shadow. Knowledge, the juggernaut! There is death in our own ritual. We must kill—dissect—destroy—and all for the sake of discovery—the worship of the ineffable light. The goddess Science demands it. We test a doubtful poison by killing. How else? No thought for self—just knowledge—the effect must be known.

The Horror in the Museum and Other Revisions
The Last Test (p. 215)

Mayo, William J.
It is a great thing to make scientific discoveries of rare value, but it is even greater to be willing to share these discoveries and to encourage other workers in the same field of scientific research.

Proceedings of Staff Meetings, Mayo Clinic
Remarks on the Romance of Medicine
Volume 10, June 19, 1935

Newton, Sir Isaac
No great discovery was ever made without a bold guess.

In W.I.B. Beveridge
The Art of Scientific Investigation
Scientist (p. 145)

Planck, Max
Scientific discovery and scientific knowledge have been achieved only by those who have gone in pursuit of it without any practical purpose whatsoever in view.

Where is Science Going?
Causation and Free Will (p. 138)

Polanyi, M.

Discoveries made by the surprising configuration of existing theories might in fact be likened to the feat of a Columbus whose genius lay in taking literally and as a guide to action that the earth was round, which his contemporaries held vaguely and as a mere matter for speculation.

In A.C. Crombie (Editor)
Scientific Change
Commentaries (p. 379)

Reichenbach, Hans

The scientist who discovers a theory is usually guided to his discovery by guesses; he cannot name a method by means of which he found the theory and can only say that it appeared plausible to him, that he had the right hunch or that he saw intuitively which assumption would fit the facts.

The Rise of Scientific Philosophy
Chapter 14 (p. 230)

Rutherford, Ernest

It is not in the nature of things for any one man to make a sudden, violent discovery; science goes step by step and every man depends on the work of his predecessors. When you hear of a sudden unexpected discovery—a bolt from the blue, as it were—you can always be sure that it has grown up by the influence of one man or another, and it is the mutual influence which makes the enormous possibility of scientific advance. Scientists are not dependent on the ideas of a single man, but on the combined wisdom of thousands of men, all thinking of the same problem and each doing his little bit to add to the great structure of knowledge which is gradually being erected.

In Robert B. Heywood
The Works of the Mind
The Scientist (p. 178)

Safonov, V.

There are scientists who make their chief discovery at the threshold of their scientific career, and spend the rest of their lives substantiating and elaborating it, mapping out the details of their discovery, as it were. There are other scientists who have to tread a long, difficult and often tortuous path to its end before they succeed in crowning their efforts with a discovery.

Courage
Chapter 10 (p. 40)

Schiller, F.C.S.

One curious result of this inertia, which deserves to rank among the fundamental 'laws' of nature, is that when a discovery has finally won tardy recognition it is usually found to have been anticipated, often with cogent reason and in great detail.

In Charles Singer (Editor)
Studies in the History and Method of Science
Volume I
Scientific Discovery and Logical Proof (pp. 256–7)

Selye, Hans
It is not to see something first, but to establish solid connections between the previously known and the hitherto unknown that constitutes the essence of scientific discovery.

From Dream to Discovery
What Should Be Done (p. 89)

Simon, Herbert
. . . scientific discovery, when viewed in detail, is an excruciatingly slow and painful process.

In Robert G. Colodny
Mind and Cosmos
Scientific Discovery and the Psychology of Problem Solving (p. 24)

Smith, Theobald
Great discoveries which give a new direction to currents of thoughts and research are not, as a rule, gained by the accumulation of vast quantities of figures and statistics. These are apt to stifle and asphyxiate and they usually follow rather than precede discovery. The great discoveries are due to the eruption of genius into a closely related field, and the transfer of the precious knowledge there found to his own domain.

Boston Medical and Surgical Journal
Volume 172, 1915 (p. 121)

Discovery should come as an adventure rather than as the result of a logical process of thought. Sharp, prolonged thinking is necessary that we may keep on the chosen road, but it does not necessarily lead to discovery.

In W.I.B. Beveridge
The Art of Scientific Investigation
Reason (p. 81)

Szent-Györgyi, Albert
Discovery consists of seeing what everybody has seen and thinking what nobody has thought.

Bioenergetics
Part II (p. 57)

Thomson, J.J.

A great discovery is not a terminus, but an avenue leading to regions hitherto unknown. We climb to the top of the peak and find that it reveals to us another higher than any we have yet seen, and so it goes on. The additions to our knowledge of physics made in a generation do not get smaller or less fundamental or less revolutionary, as one generation succeeds another. The sum of our knowledge is not like what mathematicians call a convergent series . . . where the study of a few terms may give the general properties of the whole.

In Sir George Thomson
The Inspiration of Science
Some Conclusions (p. 138)

Twain, Mark

What is it that confers the noblest delight? What is that which swells a man's breast with pride above that which any other experience can bring to him? Discovery! To know that you are walking where none others have walked; that you are beholding what human eye has not see before; that you are breathing a virgin atmosphere. To give birth to an idea—an intellectual nugget, right under the dust of a field that many a brain-plow had gone over before. To be the first—that is the idea. To do something, say something, see something, before anybody else—these are the things that confer a pleasure compared with which other pleasures are tame and commonplace, other ecstasies cheap and trivial. Lifetimes of ecstasy crowded into a single moment.

The Innocents Abroad
Chapter XXVI (p. 266)

If there wasn't anything to find out, it would be dull. Even trying to find out and not finding out is just as interesting as trying to find out and finding out; and I don't know but more so.

The Diaries of Adam and Eve
Eve's Diary
Friday (p. 87)

Unknown

Show me the scientific man who never made a mistake, and I will show you one who never made a discovery.

In J.A. Thomson
Introduction to Science
Chapter III (p. 73)

Valentine, Alan

Whenever science makes a discovery, the devil grabs it while the angels are debating the best way to use it.

Quoted in *Reader's Digest*
April 1962

Whewell, William

Advances in knowledge are not commonly made without the previous exercise of some boldness and license in guessing. The discovery of new truths requires, undoubtedly, minds careful and fertile in examining what is suggested; but it requires, no less, such as are quick and fertile in suggesting.

History of the Inductive Sciences, From the Earliest to the Present Time
Volume 1 (p. 318)

Willstätter, Richard

Whether we deal with such tentative explanations, or with the controversial protein nature of enzymes, I feel that it is not important for the scientist whether his own theory proves the right one in the end. Our experiments are not carried out to decide whether we are right, but to gain new knowledge. It is for knowledge's sake that we plow and sow. It is not inglorious at all to have erred in theories and hypotheses. Our hypotheses are intended for the present rather than for the future. They are indispensable to us in the explanation of the secured facts, to enliven and mobilize them and above all to blaze a trail into unknown regions toward new discoveries.

From My Life
Chapter 12 (p. 385)
Willard Gibbs Medal address
American Chemical Society
Chicago, September 14, 1933

DOGMA

Born, Max
When a scientific theory is firmly established and confirmed, it changes its character and becomes a part of the metaphysical background of the age: a doctrine is transformed into a dogma.

Natural Philosophy of Cause and Chance
Chapter VI (p. 47)

Eliot, George
Science is properly more scrupulous than dogma. Dogma gives a character to mistake, but the very breath of science is a contest with mistake, and must keep the conscience alive.

Middlemarch
Book VIII, LXXIII

Landheer, Barth
Science is contrasted with dogma in that it is ready to make its presuppositions the object of criticism and, if necessary, to revise them. It is not the lack of presuppositions that is the characteristic thing of science, but the self-criticism to which its principles may be subjected.

American Journal of Sociology
Presupposition in the Social Sciences (p. 544)
Volume XXXVII, January 1932

Oppenheimer, J. Robert
There must be no barriers to freedom of inquiry. There is no place for dogma in science. The scientist is free, and must be free to ask any question, to doubt any assertion, to seek for any evidence, to correct any error.

The Open Mind
The Encouragement of Science (p. 114)

Payne-Gaposchkin, Cecilia
Science is a living thing, not a dead dogma.

An Autobiography and Other Recollections
Chapter 23 (p. 233)

Pearson, Karl
Science cannot give its consent to man's development being some day again checked by the barriers which dogma and myth are ever erecting round territory that science has not yet effectively occupied.

The Grammar of Science
Introductory
Section 8 (pp. 26–7)

. . . Those who do not know the torment of the unknown cannot have the joy of discovery.
Claude Bernard – (See p. 35)

ERROR

Darwin, Charles
False facts are highly injurious to the progress of science, for they often long endure; but false views, if supported by some evidence, do little harm, as every one takes a salutary pleasure in proving their falseness.

The Descent of Man
Chapter 21

To kill an error is as good a service as, and sometimes even better than, the establishing of a new truth or fact.

More Letters of Charles Darwin
Volume II
To Wilson
March 5, 1879 (p. 422)

Huxley, Thomas
It sounds paradoxical to say the attainment of scientific truth has been effected, to a great extent, by the help of scientific errors.

Collected Essays
Volume I
Methods and Results
The Progress of Science (p. 63)

Nicolle, Charles
Error is all around us and creeps in at the least opportunity. Every method is imperfect.

In W.I.B. Beveridge
The Art of Scientific Investigation
Difficulties (p. 102)

Popper, Karl R.
But science is one of the very few human activities—perhaps the only one—in which errors are systematically criticized and fairly often, in time, corrected. This is why we can say that, in science, we often learn

from our mistakes, and why we can speak clearly and sensibly about making progress there.

Conjectures and Refutations
Chapter 10, Section I (p. 216)

Sterne, Laurence

In a word, he would say, error was error,—no matter where it fell,—whether in a fraction,—or a pound,—'twas alike fatal to truth, and she was kept down at the bottom of her well, as inevitably by mistake in the dust of a butterfly's win,—as in the disk of the sun, the moon, and all the stars of heaven put together.

Tristram Shandy
Book II, Chapter 19 (p. 150)

The laws of nature will defend themselves;—but error—(he would add, looking earnestly at my mother)—error, Sir, creeps in thro' the minute holes and small crevices which human nature leaves unguarded.

Tristram Shandy
Book II, Chapter 19 (p. 151)

Tupper, Martin Farquhar

Error is a hardy plant; it flourisheth in every coil;
In the heart of the wise and good, alike with the wicked and foolish;
For there is no error so crooked, but it hath in it some lines of truth.

Proverbial Philosophy
Of Truth in Things False

Whitehead, Alfred North

The results of science are never quite true. By a healthy independence of thought perhaps we sometimes avoid adding other people's errors to our own.

The Aims of Education
Chapter X (p. 233)

ETHICS

Baruch, Bernard M.

Science has taught us how to put the atom to work. But to make it work for good instead of for evil lies in the domain dealing with the principles of human duty. We are now facing a problem more of ethics than physics.

Address to United Nations Atomic Energy Commission
UN Headquarters
New York City
June 14, 1946

Cabot, Richard Clarke

Ethics and Science need to shake hands.

The Meaning of Right and Wrong
Introduction (p. 10)

Einstein, Albert

Scientific statements of facts and relations, indeed, cannot produce ethical derivatives. However, ethical derivatives can be made rational and coherent by logical thinking and empirical knowledge. If we can agree on some fundamental ethical propositions, then other ethical propositions can be derived from them, provided that the original premises are stated with sufficient precision. Such ethical premises play a similar role in ethics to that played by axioms in mathematics.

In Philipp Frank
Relativity—A Richer Truth
The Laws of Science and the Laws of Ethics (p. 9)

Graham, Loren R.

Science should submit to ethics, not ethics to science.

Between Science and Values
Introduction (p. 26)

Pascal, Blaise

Physical science will not console me for the ignorance of morality in the times of affliction. But the science of ethics will always console me for the ignorance of the physical sciences.

Pensées
Section II, Number 67

Russell, Bertrand

Science can discuss the causes of desires, and the means for realizing them, but it cannot contain any genuinely ethical sentences, because it is concerned with what is true or false.

Religion and Science
Science and Ethics (p. 237)

Science, by itself, cannot supply us with an ethic. It can show us how to achieve a given end, and it may show us that some ends cannot be achieved.

The New York Times Magazine
The Science to Save Us from Science
March 19, 1950

Sigma Xi

Whether or not you agree that trimming and cooking are likely to lead on to downright forgery, there is little to support the argument that trimming and cooking are less reprehensible and more forgivable. Whatever the rationalization is, in the last analysis one can no more be a little bit dishonest than one can be a little bit pregnant. Commit any of these three sins and your scientific research career is in jeopardy and deserves to be.

Honor in Science
Chapter 3 (p. 14)

Stackman, Elvin

Science cannot stop while ethics catches up . . . and nobody should expect scientists to do all the thinking for the country.

Life
U.S. Science Holds Its Biggest Powwow (p. 17)
January 9, 1950

EXPERIENCE

Bohn, H.G.
Experience is the mother of science.

A Handbook of Proverbs (p. 352)

Bohr, Niels
The great extension of our experience in recent years has brought to light the insufficiency of our simple mechanical conceptions and, as a consequence, has shaken the foundation on which the customary interpretation of observation was based . . .

Atomic Theory and the Description of Nature
Introductory Survey (p. 2)

Braithwaite, Richard B.
The peaks of science may appear to be floating in the clouds, but their foundations are in the hard facts of experience.

Scientific Explanation
Chapter XI (p. 354)

Bronowski, Jacob
Science is nothing else than the search to discover unity in the wild variety of nature—or more exactly, in the variety of our experience.

Science and Human Values
The Creative Mind (p. 27)

Cahier, Charles
La experiencia madre es de la ciencia.
[Experience is the mother of science.]

Quelques Six Mille Proverbes (p. 248)

Cervantes, Miguel de
. . . experience . . . the mother of all sciences.

Don Quixote
Part I, Book III, Chapter 7 (p. 140)

Dewey, John

It is sometimes contended, for example, that since experience is a late comer in the history of our solar system and planet, and since these occupy a trivial place in the wide area of celestial space, experience is at most a slight and insignificant incident in nature. No one with an honest respect for scientific conclusions can deny that experience is something that occurs only under highly specialized conditions, such as are found in a highly organized creature which in turn requires a specialized environment. There is no evidence that experience occurs everywhere and everywhen. But candid regard for scientific inquiry also compels the recognition that when experience does occur, no matter at what limited portion of time and space, it enters into possession of some portion of nature and in such manner as to render other of its precincts accessible.

In Austin L. Porterfield
Creative Factors in Scientific Research
Chapter II (p. 11)

Dingle, Herbert

Science . . . is the recording, augmentation, and correlation of our common human experience.

In Edgar J. Goodspeed
The Four Pillars of Democracy
Chapter II (p. 26)

Eddington, Sir Arthur Stanley

Science aims at constructing a world which shall be symbolic of the world of commonplace experience.

The Nature of the Physical World
Introduction (p. xiii)

The cleavage between the scientific and the extra-scientific domain of experience is, I believe, not a cleavage between the concrete and the transcendental, but between the metrical and non-metrical.

The Nature of the Physical World
Chapter XIII (p. 275)

Hume, David

. . . all inferences from experience suppose, as their foundation, that the future will resemble the past, and that similar powers will be cojoined with similar sensible qualities. If there be any suspicion that the course of nature may change, and that the past may be no rule for the future, all experience becomes useless, and can give rise to no inference conclusion.

Concerning Human Understanding
Section IV, Part II, 32

James, William

. . . all the magnificent achievements of mathematical and physical science—our doctrines of evolution, of uniformity of law, and the rest—proceed from our indomitable desire to cast the world into a more rational shape in our minds than the shape into which it is thrown there by the crude order of our experience.

The Will to Believe and Other Essays in Popular Philosophy
The Dilemma of Determinism (p.147)

Paré, Ambroise

Science without experience does not bring much confidence.

Attributed to one of Paré's cannons

Failure is instructive.
The person who really thinks learns quite as much from his failures as from his successes.
John Dewey – (See p. 67)

EXPERIMENT

Bacon, Francis

The present method of experiment is blind and stupid; hence men wandering and roaming without any determined course, and consulting mere chance, are hurried about to various points, and advance but little . . .

Novum Organum
First Book, 70

Experiments that yield Light are more worth while than experiments that yield Fruit.

Novum Organum
XCIX

All true and fruitful natural philosophy hath a double scale or ladder, ascendent and descendent, ascending from experiments to the invention of causes, and descending from causes to the invention of new experiments.

The Advancement of Learning
II

Bernard, Claude

The more complex the science, the more essential is it, in fact, to establish a good experimental standard, so as to secure comparable facts, free from sources of error.

An Introduction to the Study of Experimental Medicine
Introduction (p. 2)

Beveridge, W.I.B.

. . . no one believes an hypothesis except its originator but everyone believes an experiment except the experimenter.

The Art of Scientific Investigation
Chapter Four (p. 47)

Bloch, Arthur
If an experiment works, something has gone wrong.

Murphy's Law
Finagle's First Law (p. 15)

The experiment may be considered a success if no more than 50% of the observed measurements must be discarded to obtain a correspondence with the theory.

Murphy's Law
Maier's Law: Corollary (p. 47)

Bolton, Carrington
. . . reliance on the *dicta* and *data* of investigators whose very names may be unknown lies at the very foundation of physical science, and without this faith in authority the structure would fall to the ground; not the blind faith in authority of the unreasoning kind that prevailed in the Middle Ages, but the rational belief in the concurrent testimony of individuals who have recorded the results of their experiments and observations, and whose statements can be verified . . .

Quoted by J.W. Mellor in
Higher Mathematics for Students of Chemistry and Physics (p. 291)

Boyle, Robert
I neither conclude from one single Experiment, nor are the Experiments I make use of, all made upon one Subject: nor wrest I any Experiment to make it quadrare *with* any preconceiv'd Notion. But on the contrary in all kind of Experiments, and all and every one of those Trials, I make the standards (as I may say) or Touchstones by which I try all my former Notions, whether they hold not in weight or measure and touch, Ec. Foras that Body is no other than a Counterfeit Gold, which wants any one of the Properties of Gold (such as are the Malleableness, Weight, Colour, Fixtness in the Fire, Indissolubleness in Aquafortis, and the like), though it has all the other; so will all those notions be found false and deceitful, that will not undergo all the Trials and Tests made of them by Experiments. And therefore such as will not come up to the desired Apex of Perfection, I rather wholly reject and take a new, than by piercing and patching endeavour to retain the old, as knowing old things to be rather made worse by mending the better.

Quoted by Michael Roberts and E.R. Thomas in
Newton and the Origin of Colours (p. 53)

Browning, Robert
Just an experiment first, for candour's sake!

The Poems and Plays of Robert Browning
Mr. Sludge, 'The Medium'

Butler, Samuel

Every cold empirick, when his heart is expanded by a successful experiment, swells into a theorist . . .

The Works of Samuel Johnson, LL.D.
Volume II
Preface to Shakespeare (p. 340)

Carroll, Lewis

"This is the *most* interesting Experiment", the Professor announced. "It will need *time,* I'm afraid: but that is a trifling disadvantage. Now observe. If I were to unhook this weight, and let go, it would fall to the ground. You do not deny *that*?"

Nobody denied it.

"And in the same way, if I were to bend this piece of whalebone round the post—thus—and put the ring over this hook—thus—it stays bent: but, if I unhook it, it straightens itself again. You do not deny *that*?"

Again, nobody denied it.

"Well, now suppose we left things just as they are, for a long time. The force of the *whalebone* would get exhausted, you know, and it would stay bent, even when you unhooked it. Now, *why* shouldn't the same thing happen with the *weight*? The *whalebone* gets so used to being bent, that it ca'n't *straighten* itself any more. Why shouldn't the *weight* get so used to being held up, that it ca'n't *fall* any more? That's what I want to know!"

"That's what *we* want to know!" echoed the crowd.

"How long must we wait?" grumbled the Emperor.

The Professor looked at his watch. "Well, I *think* a thousand years will do to *begin* with,". . .

The Complete Works of Lewis Carroll
Sylvie and Bruno Concluded
Chapter XXI

Dalton, J.

Facts and experiments, however, relating to any subject, are never duly appreciated till, in the hand of some skilled observer, they are made the foundation of a theory by which we are able to predict the results and foresee the consequences of certain other operations which were never before undertaken. Thus a plodding experimentalist of the present time, in pursuit of the law of gravitation, might have been digging half-way to the centre of the earth in order to find the variation of gravity there . . .

In H.E. Roscoe and A. Harden
A New View of the Origin of Dalton's Atomic Theory: a Contribution to Chemical History (p. 99)

de Fontenelle, Bernard

'Tis no easy matter to be able to make an Experiment with accuracy. The least fact, which offers itself to our consideration, takes in so many other facts, which modify or compose it, that it requires the utmost dexterity to lay open the several branches of its composition, and no less sagacity to find 'em out.

Quoted by Michael Roberts and E.R. Thomas in
Newton and the Origin of Colours (p. 6)

Eddington, Sir Arthur Stanley

. . . he is an incorruptible watch-dog who will not allow anything to pass which is not observationally true.

The Philosophy of Physical Science (p. 112)

Einstein, Albert

The scientific theorist is not to be envied. For Nature, or more precisely experiment, is an inexorable and not very friendly judge of his work. It never says "Yes" to a theory. In the most favorable cases it says "Maybe," and in the great majority of cases simply "No." If an experiment agrees with a theory it means for the latter "Maybe," and if it does not agree it means "No." Probably every theory will someday experience its "No"—most theories, soon after conception.

Quoted in Helen Dukas and Banesh Hoffmann
Albert Einstein: The Human Side (p. 18)

Feynman, Richard P.

The test of all knowledge is experiment.

The Feynman Lecture on Physics
Volume I (p. 1-1)

Gay-Lussac, Joseph Louis

One cannot repeat experiments too often when the problem is one of determing a relationship.

In Maurice Grossland
Gay-Lussac: Scientist and Bourgeois
Chapter 3 (p. 70)

Godwin, William

No experiment can be more precarious than that of a half-confidence.

St. Leon; A Tale of the Sixteenth Century (p. 140)

Goethe, Johann Wolfgang von

Microscopes and telescopes, in actual fact, confuse man's innate clarity of mind.

In Ernst Lehrs
Man or Matter (p. 89)

Hume, David

. . . it being justly esteemed an unpardonable temerity to judge the whole course of nature from one single experiment, however accurate or certain.

An Enquiry Concerning Human Understanding
Section VII, 59

James, P.D.

There comes a time when every scientist, even God, has to write off an experiment.

Devices and Desires
Chapter 40

Jefferson, Thomas

. . . in the full tide of successful experiment . . .

The Inaugural Addresses of the Presidents of the United States
First Inaugural Address at Washington D.C., March 4, 1801

Kapitza, Peter Leonidovich

. . . theory is a good thing but a good experiment lasts forever.

Nature
Science East and West: Reflections of Peter Kapitza (p. 627)
(Book Review by Nevill Mott)
Volume 288, December 11, 1980

Latham, Peter Mere

Experiment is like a man traveling to some far off place, and finding no place by the way where he can sit down and rest himself, and few or no guide posts to tell him whether he be in the right direction for it or not. Still he holds on. Perhaps he has been there before, and is pretty sure of this being the direction in which he found it. Or, perhaps he has never been there, but some of his friends have, and they told him of this being the right road to it. And so it may be that, by his own sagacity and the help of well-informed friends, he reaches it at last. Or, after all his own pains, and all his friends can do for him, it may be that he never reaches it at all.

In William B. Bean
Aphorisms from Latham

Lavoisier, Antoine Laurent

We ought, in every instance, to submit our reasoning to the test of experiment, and never to search for truth but by the natural road of experiment and observation.

Elements of Chemistry
Preface

Maxwell, James Clerk

An Experiment, like every other event which takes place, is a natural phenomenon; but in a Scientific Experiment the circumstances are so arranged that the relations between a particular set of phenomena may be studied to the best advantage. In designing an Experiment the agents and the phenomena to be studied are marked off from all others and regarded as the Field of Investigation.

The Scientific Papers of James Clerk Maxwell (p. 505)

Medawar, Sir Peter

It is a truism to say that a "good" experiment is precisely that which spares us the exertion of thinking: the better it is, the less we have to worry about its interpretation, about what it "really" means.

Induction and Intuition
Chapter I (p. 15)

Pasteur, Louis

. . . this marvelous experimental method, of which one can say, in truth, not that it is sufficient for every purpose, but that it rarely leads astray, and then only those who do not use it well. It eliminates certain facts, brings forth others, interrogates nature, compels it to reply and stops only when the mind is fully satisfied. The charm of our studies, the enchantment of science, is that, everywhere and always, we can give the justification of our principles and the proof of our discoveries.

In René Dubos
Pasteur and Modern Science
Chapter I (p. 12)

Planck, Max

Experimenters are the shocktroops of science.

Scientific Autobiography (p. 110)

An experiment is a question which science poses to Nature, and a measurement is the recording of Nature's answer.

Scientific Autobiography (p. 110)

Poincaré, Henri

Experiment is the sole source of truth. It alone can teach us something new; it alone can give us certainty.

Science and Hypothesis (p. 140)

It is often said that experiments should be made without preconceived ideas. That is impossible. Not only would it make every experiment fruitless, but even if we wished to do so, it could not be done.

Science and Hypothesis (p. 143)

Rabi, Isidor I.

We don't teach our students enough of the intellectual content of experiments—their novelty and their capacity for opening new fields . . . My own view is that you take these things personally. You do an experiment because your own philosophy makes you want to know the result. It's too hard, and life is too short, to spend your time doing something because someone else has said it's important. You must feel the thing yourself . . .

The New Yorker Magazine
Profiles—Physicists, I
October 13, 1975

Rutherford, Ernest

If your experiment needs statistics, you ought to have done a better experiment.

In N.T. Bailey
The Mathematical Approach to Biology and Medicine
Chapter 2 (p. 23)

Experiment without imagination or imagination without recourse to experiment, can accomplish little, but for effective progress, a happy blend of these two powers is necessary.

Science
The Electrical Structure of Matter (p. 221)
Volume 58, 1923

Twain, Mark

It is best to prove things by actual experiment; then you *know*; whereas if you depend on guessing and supposing and conjecturing, you will never get educated.

Eve's Diary
Friday (p. 85)

Tuesday. She has taken up with a snake now. The other animals are glad, for she was always experimenting with them and bothering them; and I am glad, because the snake talks, and this enables me to get a rest.

Adam's Diary

Unknown

1. Wild enthusiasm
2. Exciting commitments
3. Total confusion
4. Re-evaluation of goals
5. Disillusionment
6. Cross-accusations
7. Search for the guilty

8. Punish the innocent
9. Promote the non-participants
10. Verbally assassinate visible leaders
11. Write and publish the report

The Eleven Phases of an Experiment

Diversity of treatment has been responsible for much of the criticism leveled against the experiment.

Source unknown

No experiment is ever a complete failure. It can always be used as a bad example.

Source unknown

You must be using the wrong equipment if an experiment works.

Source unknown

If an experiment is not worth doing at all, it is not worth doing well.

Source unknown

von Baeyer, Adolf

I never undertook my experiments to see if I was right but to see how compounds behaved. This disposition accounts for my indifference to theories.

In Richard Willstätter
From My Life
Chapter 6 (p. 140)

von Linné, Carl

I am quite aware that this road is obscured by mists that may pass over it from time to time. Yet these mists will be easily dispersed as soon as it is possible to employ widely the light of experiments. For Nature remains always the same; when she seems to be different it is because of the inevitable defects of our observations.

In Johann Wolfgang von Goethe
Botanical Writings (p. 30)

Weyl, Hermann

Allow me to express now, once and for all, my deep respect for the work of the experimenter and for his fight to wring significant facts from an inflexible Nature, who says so distinctly "No" and so indistinctly "Yes" to our theories.

The Theory of Groups and Quantum Mechanics (p. xx)

Whitehead, Alfred North

. . . experiment is nothing else than a mode of cooking the facts for the sake of exemplifying the law.

Adventures of Ideas
Foresight
Section I

There are three infallible ways of pleasing an author . . .
tell him you have read one of his books . . .
Mark Twain – (See p. 21)

FACTS

Bacon, Francis

Facts, however, will ultimately prevail; we must therefore take care that they be not against us.

In Jean Andrew de Luc
An Elementary Treatise on Geology
Section 93 (p. 82)

Chargaff, Erwin

It is not recognized sufficiently that there can be an inflation of scientific facts: the more are being produced, the less the value of each.

Perspectives in Biology and Medicine
In Praise of Smallness—How Can We Return to Small Science
Volume 23, Number 3, Spring 1980

Chesterton, G.K.

The truths of religion are unprovable; the facts of science are unproved.

The Uses of Diversity
Christian Science (p. 52)

Science itself is only the exaggeration and specialisation of this thirst for useless fact, which is the mark of the youth of man. But science has become strangely separated from the mere news and scandal of flowers and birds; men have ceased to see a pterodactyl as a pterodactyl. The rebuilding of this bridge between science and human nature is one of the greatest needs of mankind. We have all to show that before we go on to any visions or creations we can be contented with a planet of miracles.

The Apostle and the Wild Ducks
Literature and Information (p. 130)

Collingwood, R.G.

Different kinds of facts, having different degrees of scientific value, are ascertainable in these two ways. Facts ascertainable by mere observation are what are called common-sense facts, i.e. facts accessible to a commonplace mind on occasions frequent enough to be rather often perceived and of such a kind that their characteristics can be adequately perceived without trouble: so that the facts concerning them can be familiar to persons not especially gifted and not especially alert.

The New Leviathan
Part II, Chapter XXXI, aphorism 31.47

Feyerabend, Paul

. . . science is not sacrosanct. The restrictions it imposes (and there are many such restrictions though it is not easy to spell them out) are not necessary in order to have general coherent and successful views about the world. There are myths, there are the dogmas of theology, there is metaphysics, and there are many other ways of constructing a world-view. It is clear that a fruitful exchange between science and such 'non-scientific' world-views will be in even greater need of anarchism than is science itself. Thus anarchism is not only *possible*, it is *necessary* both for the internal progress of science and for the development of our culture as a whole.

Against Method
Chapter 15 (p. 180)

George, William H.

. . . the traditional way is to regard the facts of science as something like the parts of a jig-saw puzzle, which can be fitted together in one and only one way, I regard them rather as the tiny pieces of a mosaic, which can be fitted together in many ways. A new theory in an old subject is, for me, a new mosaic pattern made with the pieces taken from an older pattern.

The Scientist in Action
Personal Basis (p. 335)

Giddings, Franklin H.

The scientific study of any subject is a substitution of businesslike ways of "making sure" about it for the lazy habit of "taking it for granted" and the worse habit of making irresponsible assertions about it. To make sure, it is necessary to have done with a careless "looking into it" and to undertake precise observations, many times repeated. It is necessary to make measurements and accountings, to substitute realistic thinking (an honest dealing with facts as they are) for wishful or fanciful thinking (a self-deceiving day-dreaming) and to carry on a systematic "checking up" . . . science is nothing more nor less than getting at facts, and trying to understand them . . .

The Journal of Social Forces
Societal Variables (p. 345)
Volume I, Number 4, May 1923

Gold, Thomas

If many years go by in a field in which no significant new facts come to light, the field sharpens up the opinions and gives the appearance that the problem is solved.

Journal of Scientific Exploration
New Ideas in Science (p. 107)
Volume 3, Number 2, 1989

Hobbes, Thomas

. . . science is the knowledge of consequences, and dependence of one fact upon another . . .

Leviathan
Part I, Chapter V (p. 60)

Hoffer, Eric

The war on the present is usually a war on fact. Facts are the toys of men who live and die at leisure. They who are engrossed in the rapid realization of an extravagant hope tend to view facts as something base and unclean. Facts are counterrevolutionary.

The Passionate State of Mind
Number 73

Humboldt, Alexander von

With the simplest statements of scientific facts there must ever mingle a certain eloquence. Nature herself is sublimely eloquent. The stars as they sparkle in the firmament fill us with delight and ecstasy, and yet they all move in orbits marked out with mathematical precision.

In L. Assing (Editor)
Briefe von Alexander von Humboldt an Varnhagen von Ense
Letter
April 28, 1841

Husserl, E.

Merely fact-minded sciences make merely fact-minded people.

The Crisis of European Sciences and Transcendental Phenomenology
Part I, Section 2 (p. 6)

Koestler, Arthur

Artists treat facts as stimuli for the imagination, while scientists use their imagination to coordinate facts.

Insight and Outlook
Preface (p. vii)

Kough, A.

Facts are necessary, of course, but unless fertilized by ideas, correlated with other facts, illuminated by thought, I consider them as only material for science.

Science
The Progress of Physiology (p. 203)
Volume 70, Number 1808, August 1929

Mach, Ernst

The ultimate unintelligibilities on which science is founded must be facts, or, if they are hypotheses, must be capable of becoming facts. If the hypotheses are so chosen that their subject [*Gegenstand*] can never appeal to the senses and therefore also can never be tested, as is the case with the mechanical molecular theory, the investigator has done more than science, whose aim is facts, requires of him—and this work of supererogation is an evil.

History and Root of the Principle of the Conservation of Energy
Chapter III (p. 57)

Mayer, J.R.

If a fact is known on all its sides, it is, by that knowledge, explained, and the problem of science is ended.

Mechanik der Wärme (p. 239)

McArthur, Peter

The golden rule of science is: Make sure of your facts and then lie strenuously about your modesty.

To Be Taken with Salt (p. 150)

McCarthy, Mary

. . . in science, all facts, no matter how trivial or banal, enjoy democratic equality.

On the Contrary
The Fact in Fiction (p. 266)

Mencken, H.L.

Science, at bottom, is really anti-intellectual. It always distrusts pure reason, and demands the production of objective fact.

Minority Report: Notebook
Number 412 (p. 277)

The common view of science is that it is a sort of machine for increasing the race's store of dependable facts. It is that only in part; in even larger part it is a machine for upsetting undependable facts.

In Will Durant
Living Philosophies (p.187)

Osler, Sir William
Fed on the dry husks of facts, the human heart has a hidden want which science cannot supply . . .

Science and Immortality
The Terestans (p. 41)

Pearson, Karl
The classification of facts, the recognition of their sequence and relative significance is the function of science, and the habit of forming a judgment upon these facts unbiased by personal feeling is characteristic of what may be termed the scientific frame of mind.

The Grammar of Science
Introductory
Section 2 (p. 11)

Russell, Bertrand
A fact, in science, is not a mere fact, but an instance.

The Scientific Outlook
Chapter II (p. 59)

Sayers, Dorothy L.
. . . a false statement of fact, made deliberately, is the most serious crime a scientist can commit.

Gaudy Night
Chapter XVII (p. 340)

Schneer, Cecil J.
. . . science crossed the divide from the tidy, cultivated garden of classical thought to a new thicket of stubborn, irreducible fact.

Mind and Matter
Chapter 13 (p. 220)

Unamuno, Miguel de
Science is the most intimate school of resignation and humility, for it teaches us to bow for the seemingly most insignificant of facts.

The Tragic Sense of Life
Faith, Hope, and Charity (p. 197)

. . . science robs men of wisdom and usually converts them into phantom beings loaded up with facts.

Essays and Soliloquies
Some Arbitrary Reflections Upon Europeanization (p. 55)

Valéry, Paul
A faultily observed fact is more treacherous than a faulty train of reasoning.

The Collected Works of Paul Valéry
Volume 14
Moralités
Analectes (p. 191)

Whyte, Lancelot Law
The true aim of science is to discover a simple theory which is necessary and sufficient to cover the facts, when they have been purified of traditional prejudices.

Accent on Form
Chapter IV (p. 59)

Science does not begin with facts; one of its tasks is to uncover the facts by removing misconceptions.

Accent on Form
Chapter IV (p. 60)

Wittgenstein, Ludwig
The world is the totality of facts, not of things.

Tractatus Logico-Philosophicus
1.1

FAILURE

Dewey, John

Failure is instructive. The person who really thinks learns quite as much from his failures as from his successes.

In J. Gallian
Contemporary Abstract Algebra

Rossman, Joseph

One seldom perfects an idea without many failures . . .

Industrial Creativity: The Psychology of the Inventor
Chapter IV (p. 45)

Starling, E.H.

Every discovery, however important and apparently epoch-making, is but the natural and inevitable outcome of a vast mass of work, involving many failures, by a host of different observers, so that if it is not made by Brown this year it will fall into the lap of Jones, or of Jones and Robinson simultaneously, next year or the year after.

Nature
Discovery and Research (p. 606)
Volume 113, Number 2843, April 1924

Unknown

What is the opposite of "Eureka"?

Source unknown

FORMULA

Heisenberg, Werner

When one reduces experimental results to formalized expressions . . . thereby reaching a phenomenological description of the event, one has a feeling of having oneself invented these formulae.

Der Teil und das Ganze
Chapter 5

Kipling, Rudyard

No proposition Euclid wrote
  No formulae the text-books know,
Will turn the bullet from your coat,
  Or ward the tulwar's downward blow.
Strike hard who cares—shoot straight who can—
The odds are on the cheaper man.

Rudyard Kipling's Verse: Inclusive Edition
Arithmetic on the Frontier (p. 45)

Mitchell, Maria

Every formula which expresses a law of nature is a hymn of praise to God.

Quoted as inscription on her bust in the Hall of Fame

Peirce, Charles Sanders

It is terrible to see how a single unclear idea, a single formula without meaning, lurking in a young man's head, will sometimes act like an obstruction of inert matter in an artery, hindering the nutrition of the brain, and condemning its victim to pine away in the fullness of his intellectual vigor and in the midst of intellectual plenty.

Chance, Love, and Logic
Part I, Second Paper (p. 37)

Saint Augustine
If I am given a formula, and I am ignorant of its meaning, it cannot teach me anything, but if I already know it what does the formula teach me?

De Magistro
Chapter X, 23

What is the opposite of "Eureka"?
Unknown – (See p. 67)

FUNDING

Dunlap, Knight
It is easier for a man to get funds for what he proposes to do than for what he is doing.

Science
The Outlook for Psychology (p. 206)
Volume LXIX, Number 1782, February 22, 1929

Loehle, Craig
What would have happened if Darwin and Einstein as young men had needed to apply for government support? Their probability of getting past the grant reviewers would be similar to a snowball surviving in Hell.

BioScience
A Guide to Increased Creativity in Research—Inspiration or Perspiration?
(Figure 2, p. 125)
Volume 40, Number 2, February 1990

GOD

Bonner, W.B.
It seems to me highly improper to introduce God to solve our scientific problems.

Quoted by Charles-Albert Reichen in
A History of Astronomy (p. 100)

Compton, Karl Taylor
As the complexity of the structure of matter became revealed through research, its basic simplicity, unity, and dependability became equally evident. So we now see ourselves in a world governed by natural laws instead of by capricious deities and devils. This does not necessarily mean that God has been ruled out of the picture, but it does mean that the architect and engineer of the universe is a far different type of being from the gods assumed by the ancients, and that man lives and dies in a world of logical system and orderly performance.

A Scientist Speaks (p. 3)

Feynman, Richard P.
God was invented to explain mystery. God is always invented to explain those things that you do not understand. Now, when you finally discover how something works, you get some laws which you're taking away from God; you don't need him anymore. But you need him for the other mysteries. So therefore you leave him to create the universe because we haven't figured that out yet; you need him for understanding those things which you don't believe the laws will explain, such as consciousness, or why you only live to a certain length of time—life and death—stuff like that. God is always associated with those things that you do not understand. Therefore I don't think that the laws can be considered to be like God because they have been figured out.

In P.C.W. Davies and J. Brown
Superstrings: A Theory of Everything (p. 208)

Goodspeed, Edgar J.

. . . science is seen to be just one more of those great flights of altar stairs that lead through darkness up to God.

The Four Pillars of Democracy
Chapter II (p. 55)

Hamilton, Sir William Rowan

The genuine spirit of Mathesis is devout. No intellectual pursuit more truly leads to profound impressions of the existence and attributes of a Creator, and to a deep sense of our filial relations to him, than the study of these abstract sciences. Who can understand so well how feeble are our conceptions of Almighty Power, as he who has calculated the attraction of the sun and the planets, and weighed in his balance the irresistible force of the lightning? Who can so well understand how confused is our estimate of the Eternal Wisdom, as he who has traced out the secret laws which guide the hosts of heaven, and combine the atoms on earth? Who can so well understand that man is made in the image of his Creator, as he who has sought to frame new laws and conditions to govern imaginary worlds, and found his own thoughts similar to those on which his Creator has acted?

North American Review
The Imagination in Mathematics (pp. 226–7)
Volume 85, Number 176, July 1857

Hawking, Stephen

We still believe that the universe should be logical and beautiful; we just dropped the word "God."

In Renée Weber
Dialogues with Scientists and Sages (p. 21)

Up to now, most scientists have been too occupied with the development of new theories that describe what the universe is to ask the question why . . . If we find the answer to that, it would be the ultimate triumph of human reason—for then we would know the mind of God.

A Brief History of Time
Conclusion (p. 174)

Infeld, Leopold

Einstein uses his concept of God more often than a Catholic priest.

Quest—An Autobiography (p. 268)

James, William

The God whom science recognizes must be a God of universal laws exclusively, a God who does a wholesale, not a retail business. He cannot accommodate his processes to the convenience of individuals.

The Varieties of Religious Experience
Lecture XX (pp. 483–5)

Weil, Simone

A science which does not bring us nearer to God is worthless.

Gravity and Grace
Illusions (p. 105)

Whenever science makes a discovery, the devil grabs it while the angels are debating the best way to use it.
Alan Valentine – (See p. 41)

HISTORY OF SCIENCE

Appleton, Sir Edward

. . . the history of science has proved that fundamental research is the lifeblood of individual progress and that the ideas which lead to spectacular advances spring from it.

In J. Edwin Holmstrom
Records and Research in Engineering and Industrial Science
Chapter One (p. 7)

Bernal, J.D.

The whole history of modern science, has been that of a struggle between ideas derived from observation and practice, and pre-conceptions derived from religious training. It was not . . . that Science had to fight an external enemy, the Church; it was that the Church itself—its dogmas, its whole way of conceiving the universe—was within the scientists themselves . . . After Newton, God ruled the visible world by means of Immutable Laws of Nature, set in action by one creative impulse, but He ruled the moral world by means of absolute intimations of moral sanctions, implanted in each individual soul, reinforced and illuminated by Revelation and the Church . . . The role of God in the material world has been reduced stage by stage with the advance of Science, so much so that He only survives in the vaguest mathematical form in the minds of older physicists and biologists.

In W.H. Waddington
Science and Ethics
A Marxist Critique (pp. 115–6)

Conant, James Bryant

The history of science demonstrates beyond doubt that the really revolutionary and significant advances come not from empiricism but from new theories.

Modern Science and Modern Man
Science and Technology (p. 30)

We can put it down as one of the principles learned from the history of science that a theory is only overthrown by a better theory, never merely by contradictory facts.

On Understanding Science
The Spring of the Air (p. 36)

Feyerabend, Paul
The history of science, after all, does not just consist of facts and conclusions drawn from facts. It also contains ideas, interpretations of facts, problems created by conflicting interpretations, mistakes, and so on. On closer analysis we even find that science knows no 'bare facts' at all but that the 'facts' that enter our knowledge are already viewed in a certain way and are, therefore, essentially ideational.

Against Method
Introduction (p. 19)

Hall, A.R.
The cumulative growth of science, arising from the employment of methods of investigation and reasoning which have been justified by their fruits and their resistance to the corrosion of criticism, cannot be reduced to any single themes. We cannot say . . . why some men can perceive the truth, or a technical trick, which has eluded others. From the bewildering variety of experience in its social, economic and psychological aspects it is possible to extract only a few factors, here and there, which have had a bearing on the development of science. At present at least, we can only describe, and begin to analyse, where we should like to understand. The difficulty is the greater because the history of science is not, and cannot be, a tight unity. The different branches of science are themselves unlike in complexity, in techniques, and in their philosophy. They are not all affected equally, or at the same time, by the same historical factors, whether internal or external. It is not even possible to trace the development of a single scientific method, some formulation of principles and rules of operating which might be imagined as applicable to every scientific inquiry, for there is no such thing.

The Scientific Revolution, 1500–1800
Introduction (p. xiv)

Holton, Gerald
And yet, on looking into the history of science, one is overwhelmed by evidence that all too often there is no regular procedure, no logical system of discovery, no simple, continuous development. The process of discovery has been as varied as the temperament of the scientist.

Thematic Origins of Scientific Thought
Chapter 11 (pp. 384–5)

Huxley, Thomas

. . . any one acquainted with the history of science will admit that its progress has meant, in all ages and now more than ever, the extension of the province of *matter* and *causation*, and the gradual banishment from human thought of what we call spirit and spontaneity.

Collected Essays
Volume I
Methods and Results
On The Physical Basis of Life (p. 159)

Kuhn, Thomas S.

Though the gap seems small, there is no chasm that more needs bridging than that between the historian of ideas and the historian of science.

International Encyclopedia of the Social Sciences
Volume 14
History of Science (p. 78)

Lakatos, I.

Philosophy of science without history of science is empty; history of science without philosophy of science is blind.

In R. Buck and R. Cohen (Editors)
Boston Studies in the Philosophy of Science
Volume 8
History of Science and Rational Reconstructions (p. 91)

Mach, Ernst

We shall recognize also that not only a knowledge of the ideas that have been accepted and cultivated by subsequent teachers is necessary for the historical understanding of a science, but also that the rejected and transient thoughts of the inquirers, nay even apparently erroneous notions, may be very important and very instructive. The historical investigation of the development of a science is most needful, lest the principles treasured up in it become a system of half-understood prescripts, or worse, a system of *prejudices*. Historical investigation not only promotes the understanding of that which now is, but also brings new possibilities before us, by showing that which exists to be in great measure *conventional* and *accidental*. From the higher point of view at which different paths of thought converge we may look about us with freer vision and discover routes before unknown.

The Science of Mechanics
The Principles of Dynamics (p. 316)

Popper, Karl R.

The history of science, like the history of all human ideas, is a history of irresponsible dreams, of obstinacy, and of error.

Conjectures and Refutations
Chapter 10, Section I (p. 216)

Priestley, Joseph

The history of science cannot but animate us in our attempts to advance still further, and suggest methods and experiments to assist us in our future progress.

Quoted by John G. McEvoy
The British Journal of the History of Science
Electricity, Knowledge, and the Nature of Progress
in Priestley's Thought (p. 6)
Volume XII, Number 40, March 1979

Rand, Ayn

The entire history of science is a progression of exploded fallacies, not of achievements.

Atlas Shrugged
Part II, Chapter I (p. 337)

Richet, Charles

In the history of science, nobody has left his mark on the world unless he has been, in this sense, an innovator.

The Natural History of a Savant
Chapter VI (p. 38)

Sarton, George

From the point of view of the history of science, transmission is as essential as discovery.

Introduction to the History of Science
Volume I
Introductory Chapter (p. 15)

The only way of humanizing scientific labour is to inject into it a little of the historical spirit, the spirit of reverence for the past—the spirit of reverence for every witness of good-will through the ages. However abstract science may become, it is essentially human in its origin and growth. Each scientific result is a fruit of humanity, a proof of its virtue. The almost unconceivable immensity of the universe revealed by his own efforts does not dwarf man except in a purely physical way; it gives a deeper meaning to his life and thought. Each time that we understand the world a little better, we are also able to appreciate more keenly our relationship to it. There are no natural sciences as opposed to humanities; every branch of science or learning is just as natural or as humane as you make it. Show the deep human interest of science and the study of it becomes the best vehicle of humanism one could devise; exclude that interest, teach scientific knowledge only for the sake of information and professional instruction, and the study of it, however valuable from a purely technical point of view, loses all educational value. Without history, scientific knowledge may become culturally dangerous; combined with history, tempered with reverence, it will nourish the highest culture.

History of Science and the New Humanism (pp. 68–9)

Schweizer, Karl W.

One of the obstructions to a genuine appreciation of history is the existence of a vague unformulated assumption that historical research merely seeks to disinter a fossilized past—merely digs into the memory to recover things which the human race once knew before. On the basis of such an assumption it is possible for people to have the feeling that history can never produce anything which is fundamentally novel, but merely fills our minds with the lumber of bygone ages.

Herbert Butterfield: Essays on the History of Science
Chapter II (p. 19)

Tannery, Paul

The scientist in so far as he is a scientist is only drawn to the history of the particular science that he studies himself; he will demand that this history be written with every possible technical detail, for it is only thus that it can supply him with materials of any possible utility. But what he will particularly require is the study of the thread of ideas and the linking together of discoveries. *His* chief object is to rediscover in its original form the expression of his predecessors' actual thoughts, in order to compare them with his own; and to unravel the methods that served in the construction of current theories, in order to discover at what point and towards what goal an effort towards innovation may be made.

In A. Rupert Hall
The British Journal for the History of Science
Can the History of Science be History? (p. 212)
Volume IV, Part III, Number 15, June 1969

Whewell, William

It will be universally expected that a history of Inductive Science should . . . afford us some indication of the most promising mode of directing our future efforts to add to its extent and completeness.

History of the Inductive Sciences, From the Earliest to the Present Time
Volume I
Introduction (p. 5)

. . . the existence of clear Ideas applied to distinct Facts will be discernible in the History of Science, whenever any marked advance takes place. And in tracing the progress of the various provinces of knowledge which come under our survey, it will be important for us to see that, at all such epochs, such a combination has occurred . . .

In our history, it is the *progress* of knowledge only which we have to attend to. This is the main action of our drama; and all the events which do not bear upon this, though they may relate to the cultivation and the cultivators of philosophy, are not a necessary part of our theme.

History of the Inductive Sciences, From the Earliest to the Present Time
Volume I
Introduction (pp. 9, 12)

Whitehead, Alfred North
Science is concerned with the facts of bygone transition. History relates the aim at ideals. And between Science and History, lies the operation of the Deistic impulse of energy. It is the religious impulse in the world which transforms the dead facts of Science into the living drama of History. For this reason Science can never foretell the perpetual novelty of History.

Modes of Thought
Chapter Five (p. 142)

Williams, L. Pearce
. . . the history of science is a professional and rigorous discipline demanding the same level of skills and scholarship as any other scholarly field. It is time for the scientist to realize that he studies nature and others study him. He is no more nor no less competent to comment on his own activities and the activities of his fellow scientist than is the politician. Critical political history is rarely written by the politician and the same is true of the history of science.

Scientific American
Letter to the Editor (p. 8)
Volume 214, Number 6, June 1966

HYPOTHESIS

Bartlett, Elisha
The restless and inquisitive mind, from its very constitution insatiable, and ever unsatisfied with its actual and absolute possessions, endeavors to imagine the phenomena, which it cannot demonstrate; it struggles to overleap the boundary, whose inexorable circumference cages it in; and, failing to do this, it fills the infinite and unknown regions, beyond and without it, with its own creations.

An Essay on the Philosophy of Medical Science
Part I, Chapter 4 (p. 33)

Beveridge, W.I.B.
The hypothesis is the principal intellectual instrument in research. Its function is to indicate new experiments and observations and it therefore sometimes leads to discoveries even when not correct itself.

We must resist the temptation to become too attached to our hypothesis, and strive to judge it objectively and modify it or discard it as soon as contrary evidence is brought to light. Vigilance is needed to prevent our observations and interpretations being biased in favor of the hypothesis. Suppositions can be used without being believed.

The Art of Scientific Investigation
Hypothesis (p. 52)

Boltzmann, Ludwig
. . . neither the Theory of Gases nor any other physical theory can be quite a congruent account of facts . . . Certainly, therefore, Hertz is right when he says: "The rigour of science requires, that we distinguish well the undraped figure of nature itself from the gay-coloured vesture with which we clothe it at our pleasure." But I think the predilection for nudity would be carried too far if we were to forego every hypothesis.

Lectures on Gas Theory
Translator's Introduction (p. 16)

Davy, Sir Humphrey

The only use of an hypothesis is, that it should lead to experiments; that it should be a guide to facts. In this application, conjectures are always of use. The destruction of an error hardly ever takes place without the discovery of truth . . .

Hypothesis should be considered merely an intellectual instrument of discovery, which at any time may be relinquished for a better instrument. It should never be spoken of as truth; its highest praise is verisimility. Knowledge can only be acquired by the senses; nature has an archetype in the human imagination; her empire is given only to industry and action, guided and governed by experience.

In John Davy (Editor)
Collected Works
Volume 8 (pp. 346–7)

[Hypotheses were] part of the scaffolding of the building of science [rather] than as belonging either to its foundations, materials, or ornaments.

Fragmentary Remains
Chapter VII (pp. 231–2)

de Morgan, Augustus

. . . wrong hypotheses, rightly worked from, have produced more useful results than unguided observation.

A Budget of Paradoxes
Volume I (p. 87)

Doyle, Sir Arthur Conan

If the fresh facts which come to our knowledge all fit themselves into the scheme, then our hypothesis may gradually become a solution.

The Complete Sherlock Holmes
The Adventure of Wisteria Lodge

I have devised seven separate explanations, each of which would cover the facts as far as we know them. But which of these is correct can only be determined by the fresh information which we shall no doubt find waiting for us.

The Complete Sherlock Holmes
The Adventure of the Copper Beeches (p. 511)

Friedman, Milton

The construction of hypotheses is a creative act of inspiration, intuition, invention; its essence is the vision of something new in familiar material.

Essays in Positive Economics
Part I, Section VI (p. 43)

Goldenweiser, Alexander
Scientific hypotheses are intuitive leaps in the dark.

Robots or Gods (p. 46)

Gregg, Alan
A dream is a firstborn: an hypothesis should be an orphan.

In Wilder Penfield
The Difficult Art of Giving (p. 318)

Huxley, Thomas
But the great tragedy of Science—the slaying of a beautiful hypothesis by an ugly fact . . .

Collected Essays
Volume VIII
Biogenesis and Abiogenesis
Discourses Biological and Geological (p. 244)

Lorenz, Konrad
It is a good morning exercise for a research scientists to discard a pet hypothesis every day before breakfast. It keeps him young.

On Aggression
Chapter Two (p. 12)

Maine, Sir Henry
Next to what a modern satirist has called "Hypothetics"—the science of that which might have happened but did not . . .

Popular Government: Four Essays
Essay I (p. 4)

Medawar, Sir Peter
I cannot give any scientist of any age better advice than this: the intensity of a conviction that a hypothesis is true has no bearing over whether it is true or not.

Advice to a Young Scientist
Chapter 6 (p. 39)

Newton, Sir Isaac
. . . I frame no hypotheses; for whatever is not deduced from the phenomena is to be called an hypothesis; and hypotheses, whether metaphysical or physical, whether of occult qualities or mechanical, have no place in experimental philosophy.

Mathematical Principles of Natural Philosophy
Book III, General Scholium (p. 371)

Nordmann, Charles
Hypotheses in science are a kind of soft cement which hardens rapidly in the open air, thus enabling us to join together the separate blocks of

the structure and to fill up the breaches made in the walls by projectiles, with artificial stuff which the superficial observer presently mistakes for stone. It is because hypotheses are something like that in science that the best scientific theories are those which include least hypotheses.

Einstein and the Universe
Chapter II (p. 40)

Peirce, Charles Sanders

If hypotheses are to be tried haphazard, or simply because they will suit certain phenomena, it will occupy the mathematical physicists of the world say half a century on the average to bring each theory to the test, and since the number of possible theories may go up into the trillion, only one of which can be true, we have little prospect of making further solid additions to the subject in our time.

The Monist
The Architecture of Theories (p. 164)
Volume I, Number 2, January 1891

The great difference between induction and hypothesis is, that the former infers the existence of phenomena such as we have observed in cases which are similar, while hypothesis supposes something of a different kind from what we have directly observed, and frequently something which it would be impossible for us to observe directly.

Chance, Love, and Logic
Deduction, Induction, Hypothesis (p. 149)

Planck, Max

For every hypothesis in physical science has to go through a period of difficult gestation and parturition before it can be brought out into the light of day and handed to others, ready-made in scientific form so that it will be, as it were, fool-proof in the hands of outsiders who wish to apply it.

Where is Science Going?
From Relative to Absolute (p. 178)

Popper, Karl R.

The best we can say of a hypothesis is that up to now it has been able to show its worth, and that it has been more successful than other hypotheses although, in principle, it can never be justified, verified or even shown to be probable. The appraisal of the hypothesis relies solely upon deductive consequences (predictions) which may be drawn from the hypothesis: *There is no need even to mention 'induction'.*

The Logic of Scientific Discovery
New Appendices
Two Notes on Induction and Demarcation 1933–1934 (p. 315)

Poynting, J.H.

The hypotheses of science are continually changing. Old hypotheses break down and new ones take their place. But the classification of known phenomena which a hypothesis has suggested, and the new discoveries of phenomena to which it has led, remain as positive and permanent additions to natural knowledge when the hypothesis itself has vanished from thought.

In J.A. Thomson
Introduction to Science
Chapter I (p. 27)

Priestley, Joseph

Hypotheses, while they are considered merely as such, lead persons to try a variety of experiments, in order to ascertain them. In these experiments, new facts generally arise. These new facts serve to correct the hypothesis which gave occasion to them. The theory, thus corrected, serves to discover more new facts, which, as before, bring the theory still nearer to the truth. In this progressive state, or method of approximation, things continue . . .

The History and Present State of Electricity
Volume II
Part III, Section I (pp. 15–6)

Richet, Charles

Be as bold in the conception of hypotheses as rigorous in their demonstration.

The Natural History of a Savant
Chapter X (p. 123)

Sterne, Laurence

It is the nature of an hypothesis, when once a man has conceived it, that it assimilates every thing to itself, as proper nourishment; and, from the first moment of your begetting it, it generally grows the stronger by every thing you see, hear, read, or understand. This is of great use.

Tristram Shandy
Book 2, Chapter 19 (p. 156)

Weismann, August

. . . when we are confronted with facts which we see no possibility of understanding save on a single hypothesis, even though it be an undemonstratable one, we are naturally led to accept the hypothesis, at least until a better one can be found.

The Evolution Theory
Volume I
Lecture XII (p. 242)

IMAGINATION

Brodie, Sir Benjamin
Lastly, physical investigation, more than anything besides, helps to teach us the actual value and right use of the Imagination—of that wondrous faculty, which, left to ramble uncontrolled, leads us astray into a wilderness of perplexities and errors, a land of mists and shadows; but which, properly controlled by experience and reflection, becomes the noblest attribute of man; the source of poetic genius, the instrument of discovery in Science, without the aid of which Newton would never have invented fluxions, nor Davy have composed the earths and alkalies, nor would Columbus have found another Continent.

Address to the Royal Society
November 30, 1859
In John Tyndall
Fragments of Science
Scientific Use of the Imagination (p. 423)

Brougham, Henry
A mere theory . . . is the unmanly and unfruitful pleasure of a boyish and prurient imagination, or the gratification of a corrupted and depraved appetite . . .

Edinburgh Review
The Bakerian Lecture on the Theory of Light and Colours (p. 450)
Volume 1, 1801–3

Dewey, John
Every great advance in science has issued from a new audacity of imagination.

The Quest for Certainty
XI (p. 310)

Douglas, A. Vibert

To every investigator there come moments when his thought is baffled, when the limits of experimental possibility seem to have been reached and he faces a barrier which defies his curiosity. Then it is that imagination, like a glorious greyhound, comes bounding along, leaps the barrier, and a vision is flashed before the mind . . .

The Atlantic Monthly
From Atoms to Stars (p. 158)
August 1929

Einstein, Albert

Imagination is more important than knowledge. For knowledge is limited, whereas imagination embraces the entire world, stimulating progress, giving birth to evolution. It is, strictly speaking, a real factor in scientific research.

Cosmic Religion
On Science (p. 97)

Faraday, Michael

The truth of science has ever had not merely the task of evolving herself from the dull and uniform mist of ignorance, but also that of repressing and dissolving the phantoms of the imagination.

In Louis Trenchard More
The Limitations of Science
Chapter IV (p. 107)

Harding, Rosamund E.M.

Dreaming over a subject is simply . . . allowing the will to focus the mind passively on the subject so that it follows the trains of thought as they arise, stopping them only when unprofitable but in general allowing them to form and branch naturally until some useful and interesting results occur.

An Anatomy of Inspiration
Chapter I (p. 5)

Huggins, William

This creative use of the imagination is not only the fountain of all inspiration in poetry and art, but is also the source of discovery in science, and indeed supplies the initial impulse to all development and progress. It is this creative power of the imagination which has inspired and guided all the great discoveries in science.

In William H. George
The Scientist in Action
The Scientific Theory (p. 226)

Mayo, William J.

The sciences bring into play the imagination, the building of images in which the reality of the past is blended with the ideals for the future, and from the picture there springs the prescience of genius.

The Journal of the American Medical Association
Contributions of Pure Science to Progressive Medicine (p. 1466)
Volume 84 Number 20, May 16, 1925

Patten, William

Imagination opens the gates of the universe.

In Austin L. Porterfield
Creative Factors in Scientific Research
Chapter IV (p. 61)

Pearson, Karl

All great scientists have, in a certain sense, been great artists; the man with no imagination may collect facts, but he cannot make great discoveries.

Grammar of Science
Introductory
Section 11 (p. 31)

Hundreds of men have allowed their imagination to solve the universe, but the men who have contributed to our real understanding of natural phenomena have been those who were unstinted in their application of criticism to the product of their imaginations.

Grammar of Science
Introductory
Section 11 (p. 32)

Pulitzer, Joseph

I know what you mean by imagination! That it is necessarily inexact or irresponsible. I hope you will recover from that. Imagination isn't disorder or sloppiness or substituting misinformation for something that should have been definitely ascertained . . . It isn't being lazy or indifferent or lacking personal or professional conscience. No. It is what the astronomer has when he says that right there, though no one has located it, must be a star. It is what Darwin had when, with the long orchid in his hand, he said that somewhere they would find the long-tongued moth who visited it.

In Austin L. Porterfield
Creative Factors in Scientific Research
Chapter IV (p. 61)

Thomson, Sir George

Science, like all arts, needs imagination.

The Inspiration of Science
The Scientific Method (p. 8)

Tyndall, J.

With accurate experiment and observation to work upon, imagination becomes the architect of physical theory.

In W.I.B. Beveridge
The Art of Scientific Investigation
Imagination (p. 53)

Wilson, Edmund

The great scientists have been occupied with values—it is only their vulgar followers who think they are not. If scientists like Descartes, Newton, Einstein, Darwin, and Freud don't "look deeply into experience," what do they do? They have imaginations as powerful as any poet's and some of them were first-rate writers as well. How do you draw the line between *Walden* and *The Voyage of the Beagle*? The product of the scientific imagination is a new vision of relations—like that of the artistic imagination.

Letters on Literature and Politics, 1912–1972
Letter to Allen Tate, July 20, 1931 (p. 212)

INDUCTION

Comte, Auguste

Induction for deduction, with a view to construction.

In J.A. Thomson
Introduction to Science
Chapter III (p. 58)

Lewis, C.S.

This is called the inductive method. Hypothesis, my dear young friend, establishes itself by a cumulative process: or, to use popular language, if you make the same guess often enough it ceases to be a guess and becomes a Scientific Fact.

The Pilgrim's Regress: An Allegorical Apology for Christianity, Reason and Romanticism
Chapter One (p. 36)

Whitehead, Alfred North

If science is not to degenerate into a medley of *ad hoc* hypotheses, it must become philosophical and must enter upon a thorough criticism of its own foundations.

Science and the Modern World
Chapter I (p. 17)

INFORMATION

Bernal, J.D.

We should admit in theory what is already very largely a case in practice, that the main currency of scientific information is the secondary sources in the form of abstracts, reports, tables, etc., and that the primary sources are only for detailed reference by very few people. It is possible that the fate of most scientific papers will be not to be read by anyone who uses them, but with luck they will furnish an item, a number, some facts or data to such reports which may, but usually will not, lead to the original paper being consulted. This is very sad but it is the inevitable consequence of the growth of science.

Journal of Documentation
The Supply of Information to the Scientist: Some Problems of the Present Day
Volume 13, 1957

Dretske, Fred I.

Beauty is in the eye of the beholder, and information is in the head of the receiver.

Knowledge and the Flow of Information (p. vii)

Eliot, T.S.

Where is the Life we have lost in living? Where is the wisdom we have lost in knowledge? Where is the knowledge we have lost in information?

The Rock
I

Marquis, D.G.
Allen, T.J.

Far more is known about the flow of information among scientists than among technologists. From the knowledge that is available, however, we are led to conclude that the communication patterns in the two areas of activity are not only largely independent of one another, but qualitatively different in their nature. This difference is reflected most clearly in the mechanisms by which information is diffused within the two sets of practitioners.

American Psychiatrist
Communication Patterns in Applied Technology
Volume 21

President's Science Advisory Committee
We shall cope with the information explosion, in the long run, only if some scientists and engineers are prepared to commit themselves deeply to the job of sifting, reviewing, and synthesizing information; i.e., to handling information with sophistication and meaning, not merely mechanically. Such scientists must create new science, not just shuffle documents: their activities of reviewing, writing books, criticizing, and synthesizing are as much a part of science as is traditional research.

Science, Government and Information (p. 2)

Like a man on a bicycle science cannot stop;
it must progress or collapse.
E.E. Free – (See p. 154)

INSTRUMENTS

Bacon, Francis

The unassisted hand and the understanding left to itself possess but little power. Effects are produced by the means of instruments and helps, which the understanding requires no less than the hand; and as instruments either promote or regulate the motion of the hand, so those that are applied to the mind prompt or protect the understanding.

Novum Organum
First Book
2

Beveridge, W.I.B.

Elaborate apparatus plays an important part in the science of to-day, but I sometimes wonder if we are not inclined to forget that the most important instrument in research must always be the mind of man.

The Art of Scientific Investigation
Preface (p. ix)

Chesterton, G.K.

Men can construct a science with very few instruments, or with very plain instruments; but no one on earth could construct a science with unreliable instruments.

Heretics
Science and the Savages (p. 147)

Davy, Sir Humphrey

. . . nothing tends so much to the advancement of knowledge as the application of a new instrument. The native intellectual powers of men in different times are not so much the causes of the different success of their labours, as the peculiar nature of the means and artificial resources in their possession.

In Thomas Hager
Force of Nature: the Life of Linus Pauling
1 (p. 86)

Flexner, Abraham
Science lies in the intellect, not in the instruments.

Medical Education
Chapter I (pp. 6–7)

Giere, Ronald
The overwhelming presence of machines and instrumentation must be one of the most salient features of the modern scientific laboratory . . . The development of science depends at least as much on new machinery as it does on new ideas.

Explaining Science: A Cognitive Approach (p. 138)

Whitehead, Alfred North
The reason why we are on a higher imaginative level is not because we have finer imagination, but because we have better instruments.

Science and the Modern World
Chapter VII (p. 114)

INTUITION

Einstein, Albert

. . . there is no logical way to the discovery of these elemental laws. There is only the way of intuition, which is helped by a feeling for the order lying behind the appearance . . .

In Max Planck
Where is Science Going?
Prologue (p. 10)

To these elementary laws there leads no logical path, but only intuition, supported by being sympathetically in touch with experience.

In Gerald Holton
Thematic Origins of Scientific Thought
Chapter 10 (p. 357)

Huxley, Aldous

. . . experience is not a matter of having actually swum the Hellespont, or danced with the dervishes, or slept in a doss-house. It is a matter of sensibility and intuition, of seeing and hearing the significant things, of paying attention at the right moments, of understanding and co-ordinating. Experience is not what happens to a man; it is what a man does with what happens to him.

Texts and Pretexts
Introduction (p. 5)

INVENTION

Bacon, Francis

The real and legitimate goal of the sciences is the endowment of human life with new inventions and riches.

Novum Organum
Aphorism LXXXI

Middendorf, W.H.
Brown, G.T., Jr.

A full storehouse of knowledge is a necessary but not sufficient condition for invention. To this, one must add an organized method of attack.

Electrical Engineering
Orderly Creative Inventing (p. 867)
October 1957

Sprat, Thomas

Invention is an *Heroic* thing and plac'd above the reach of a low, and vulgar *Genius*. It requires an *active*, a bold, a nimble, a restless *mind*: a thousand difficulties must be contemn'd, with which a mean heart would be broken: many *attempts* must be made to no purpose: much *Treasure* must be scattered without any return: much violence and vigor of thought must attend it: some irregularities and excesses must be granted it that would hardly be pardon'd by the severe *Rules of Prudence*.

The History of the Royal Society of London for the Improving of Natural Knowledge
Section XXXI (p. 392)

IRRATIONAL

Laudan, Larry

If rationality consists in believing only what we can reasonably presume to be true, and if we define 'truth' in its classical nonpragmatic sense, then science is (and will forever remain) irrational.

Progress and its Problems
Chapter Four (p. 125)

. . . *when a thinker does what it is rational to do, we need inquire no further into the causes of his actions*; whereas, when he does what is in fact irrational—even if he believes it to be rational—we require some further explanation.

Progress and its Problems
Chapter Six (pp. 188–9)

Mannheim, Karl

Anyone who wants to drag in the irrational where the lucidity and acuity of reason still must rule by right merely shows that he is afraid to face the mystery at its legitimate place.

Essays on the Sociology of Knowledge
Chapter V (p. 229)

Wilde, Oscar

CHILTERN: You think science cannot grapple with the problem of women?

CHEVELEY: Science can never grapple with the irrational. That is why it has no future before it, in this world.

CHILTERN: And women represent the irrational.

CHEVELEY: Well-dressed women do.

An Ideal Husband
Act I

KNOWLEDGE

Aristotle
We think we have scientific knowledge when we know the cause, and there are four causes: (1) the definable form, (2) an antecedent which necessitates a consequent, (3) the efficient cause, (4) the final cause.

Posterior Analytics
Book II, Chapter 11, 94^a, [20]

Yet it does not appear to be true in all cases that correlatives come into existence simultaneously. The object of knowledge would appear to exist before knowledge itself for it is usually the case that we acquire knowledge of objects already existing; it may be difficult, if not impossible, to find a branch of knowledge the beginning of the existence of which was contemporaneous with that of its object.

Categories
Chapter 7, 7^b, [20]

Bacon, Francis
. . . that knowledge hath in it somewhat of the serpent, and therefore where it entereth into a man it makes him swell; *"scientia inflat"* [knowledge puffs up].

Advancement of Learning
First Book, Chapter I, 2

Barrie, James Matthew
ERNEST: . . . I'm not young enough to know everything.

The Admirable Crichton
Act I (p. 16)

Barry, Frederick
It is clear, for instance, that we classify business management, pugilism and medicine together as science because, though as occupations they are only incidentally related, they are all characterized by the practical,

methodical, and so far as is humanly possible, the rational utilization of knowledge for the attainment of definite goals.

The Scientific Habit of Thought
Chapter I (p. 5)

Bloch, Marc
Each science, taken by itself, represents but a fragment of the universal march toward knowledge.

The Historian's Craft (p. 18)

Bube, R.H.
Science has become a particular kind of knowledge obtained in a particular way: knowledge of the natural world obtained by sense interaction with that world.

The Encounter between Science and Christianity
Chapter 1 (p. 17)

Bush, Vannevar
The process by which the boundaries of knowledge are advanced, and the structure of organised science is built, is a complex process indeed. It corresponds fairly well with the exploitation of a difficult quarry for its building materials and the fitting of these into an edifice; but there are very significant differences. First, the material itself is exceedingly varied, hidden and overlaid with relatively worthless rubble . . . Second, the whole effort is highly unorganised. There are no direct orders from architect or quarrymaster. Individuals and small bands proceed about their business unimpeded and uncontrolled, digging where they will, working over their material, and tucking it into place in the edifice.

Endless Horizons
Chapter 17 (p. 179)

Byron, Lord George Gordon
That knowledge is not happiness, and science
But an exchange of ignorance for that
Which is another kind of ignorance.

The Works of Lord Byron
Manfred, A Dramatic Poem
Act II, Scene 4, L. 431–433 (p. 227)

Carnap, Rudolf
When we say that scientific knowledge is unlimited, we mean 'there is no question whose answer is in principle unattainable by science'.

New Scientist
In Mary Midgley
Can Science Save Its Soul? (p. 24)
August 1, 1992

Collingwood, R.G.
Questioning is the cutting edge of knowledge; assertion is the dead weight behind the edge that gives it driving force.

Speculum Mentis (p. 78)

Collins, Wilkie
. . . what is scientific knowledge now may be scientific ignorance in some years more.

Heart and Science
LIV (p. 285)

Conant, James Bryant
The stumbling way in which even the ablest of the scientists in every generation have had to fight through thickets of erroneous observations, misleading generalizations, inadequate formulations, and unconscious prejudice is rarely appreciated by those who obtain their scientific knowledge from textbooks.

Science and Common Sense
Chapter Three (p. 44)

Cowper, William
Knowledge and Wisdom, far from being one,
Have oft-times no connection. Knowledge dwells
In heads replete with thoughts of other men;
Wisdom in minds attentive to their own.
Knowledge, a rude unprofitable mass,
The mere materials with which wisdom builds . . .
Knowledge is proud that he has learn'd so much;
Wisdom is humble that he knows no more.

Complete Poetical Works
The Task
Book VI, L. 88–93, 96, 97

da Vinci, Leonardo
The acquisition of any knowledge whatever is always useful to the intellect, because it will be able to banish the useless things and retain those which are good. For nothing can be either loved or hated unless it is first known.

In Edward MacCurdy
The Notebooks of Leonardo da Vinci
Volume I
Aphorisms (p. 88)

All knowledge originates in opinions.

Leonardo da Vinci's Notebooks (p. 53)

Darwin, Charles

The more one thinks, the more one feels the hopeless immensity of man's ignorance.

More Letters of Charles Darwin
Volume I
To Farrar
August 28, 1881 (p. 394)

Drexler, K. Eric

People who confuse science with technology tend to become confused about limits . . . they imagine that new knowledge always means new know-how; some even imagine that knowing everything would let us do anything.

Engines of Creation
Chapter 10 (p. 148)

Ecclesiastes 1:18

For with much wisdom comes much sorrow; the more knowledge, the more grief.

The Bible

Egler, Frank E.

Knowledge is not wisdom; wisdom is knowledge, when it is tempered by judgment.

The Way of Science
The Nature of Science (p. 1)

Einstein, Albert

The aspect of knowledge that has not yet been laid bare gives the investigator a feeling akin to that experienced by a child who seeks to grasp the masterly way in which elders manipulate things.

In Alexander Moszkowski
Conversations with Einstein
Chapter III (p. 46)

It is my inner conviction that the development of science seeks in the main to satisfy the longing for pure knowledge.

In Alexander Moszkowski
Conversations with Einstein
Chapter VIII (p. 173)

Yet it is equally clear that knowledge of what *is*, does not open the door directly to what *should be*.

Out of My Later Years (p. 22)

Faber, Harold

New scientific knowledge is like wine in the wedding of Cana: it cannot be used up; the same idea can serve many users simultaneously; and as the number of customers increases, no one need be getting less of it because the others are getting more.

The Book of Laws
The Laws of Economics
Leontief's Law (p. 57)

Harth, Erich

I have no possessions that are truly my own. I am like a stranger at a rich man's gate. What I have is borrowed, and even my knowledge is nothing but hand-me-downs, and an occasional oddity I pick up by chance. I pass it on to others like me.

The Creative Loop: How the Brain Makes a Mind
Chapter 1 (p. 6)

Harvey, William

What I shall deliver in these my Exercises on Animal Generation I am anxious to make publicly known, not merely that posterity may there perceive the sure and obvious truth, but further, and especially, that by exhibiting the method of investigation which I have followed, I may propose to the studious a new and unless I am mistaken a safer way to the attainment of knowledge.

Anatomical Exercises on the Generation of Animals
Introduction (p. 331)

Hinshelwood, C.N.

To some men knowledge of the universe has been an end possessing in itself a value that is absolute: to others it has seemed a means of useful applications.

The Structure of Physical Chemistry
Atoms and Molecules (p. 2)

Holmes, Oliver Wendell

Scientific knowledge, even in the most modest persons, has mingled with it a something which partakes of insolence.

The Autocrat of the Breakfast-Table
Chapter 3

It is the province of knowledge to speak and it is the privilege of wisdom to listen.

The Poet at the Breakfast-Table
Chapter 10

Holton, Gerald
Roller, Duane H.D.

Knowledge is like a net; if one were to cut out the part which is labeled science, the rest of the net would be useless.

Foundations of Modern Physical Science
Chapter 14 (p. 247)

Hubble, Edwin

Science acquires knowledge but has no interest in its practical applications. The applications are the work of engineers.

The Nature of Science
Scientists at War (p. 63)

Huxley, Aldous

Knowledge is power and, by a seeming paradox, it is through their knowledge of what happens in this unexperienced world of abstractions and inferences that scientists and technologists have acquired their enormous and growing power to control, direct and modify the world of manifold appearances in which human beings are privileged and condemned to live.

Literature and Science
Chapter 3 (p. 9)

Huxley, Thomas

Indeed, if a little knowledge is dangerous, where is the man who has so much as to be out of danger?

Collected Essays
Volume III
On Elementary Instruction in Physiology

In science, as in life, learning and knowledge are distinct, and the study of things, and not of books, is the source of the latter.

Collected Essays
Volume VIII
The Study of Zoology

Huxley, Thomas
Huxley, Julian

Knowledge is not merely an end in itself, but the only satisfactory means for controlling our further evolution.

Evolution and Ethics: 1893–1943 (p. 133)

Jeans, Sir James

. . . science should leave off making pronouncements: the river of knowledge has too often turned back on itself.

The Mysterious Universe
Chapter V (p. 188)

Kettering, Charles F.

We have only begun to knock a few chips from the great quarry of knowledge that has been given us to dig out and use. We know almost nothing about everything. That is why, with all conviction, I say that the future is boundless.

In James Kip Finch
Engineering and Western Civilization
Chapter 20 (p. 306)

Larrabee, Harold A.

There are some human minds, we know, in which the yearning for a complete impersonal and consistent system of knowledge is a strong and even dominant desire. But the fact that they make this titanic demand upon the universe in no way guarantees that the latter will turn out to be amenable to such orderings, especially in terms of a single type of system.

Reliable Knowledge
Chapter 2, Section 5 (p. 72)

Lichtenberg, Georg Christoph

I made the journey to knowledge like dogs who go for walks with their masters, a hundred times forward and backward over the same territory; and when I arrived I was tired.

Lichtenberg: Aphorisms & Letters
Aphorisms (p. 58)

Loehle, Craig

A major obstacle to science is not ignorance but knowledge.

Bioscience
A Guide to Increased Creativity in Research—Inspiration or Perspiration?
Volume 40, Number 2, February 1990 (p. 123)

Mayo, William J.

Science is organized knowledge of the physical world.

Collected Papers of the Mayo Clinic & Mayo Foundation
Perception
Volume 20, 1928

Milton, John

O Sacred, Wise, and Wisdom-giving Plant,
Mother of Science . . .

Paradise Lost
Chapter IX, L. 679–680

Myrdal, Gunnar

All ignorance, like all knowledge, tends thus to be opportunist.

Objectivity in Social Research
Chapter III (p. 19)

Oparin, A.I.

One can only understand the essence of things when one knows their origin and development.

Life: Its Nature, Origin and Development
Chapter I (p. 37)

Osler, Sir William

Science is organized knowledge, and knowledge is of things we see. Now the things that are seen are temporal; of the things that are unseen science knows nothing and has at present no means of knowing anything.

Science and Immortality
The Terestans (pp. 40–1)

Popper, Karl R.

. . . we should have to represent the tree of knowledge as springing from countless roots which grow up into the air rather than down, and which ultimately, high up, tend to unite into one common stem.

Objective Knowledge
Chapter 7 (pp. 262–3)

Price, C.

As new knowledge develops, it has increasingly provided natural explanations for facts and phenomena formerly ascribed to the supernatural. Perhaps an understanding of chemical evolution and biological function can develop a philosophy of man more unified, less divisive, less of a major breeding ground for man's inhumanity to man than the many religious dogmas now so much used to inflame feelings of hatred, suspicion and prejudice in human society.

In D. Fohlfing and A. Oparin (Editors)
Molecular Evolution: Prebiological and Biological
Some Social and Philosophical Implications of Progress
on the Origin and Synthesis of Life (p. 462)

Richet, Charles

The aim of science is knowledge about phenomena.

The Natural History of a Savant
Chapter VI (p. 42)

Rossman, Joseph

However, knowledge alone, . . . never gives rise to new inventions or industries. It is usually left to the inventor to utilize the facts and principles of science, and to apply them for practical purposes.

Industrial Creativity: The Psychology of the Inventor (p. 19)

Royce, Josiah

Science is never merely knowledge; it is orderly knowledge.

Encyclopaedia of the Philosophical Sciences
Volume I
Logic
The Principles of Logic

Russell, Bertrand

Whatever knowledge we possess is either knowledge of particular facts or scientific knowledge.

The Scientific Outlook
Chapter III (p. 73)

Science, as its name implies, is primarily knowledge; by convention it is knowledge of a certain kind, the kind namely, which seeks general laws connecting a number of particular facts.

The Scientific Outlook
Introduction (p. 10)

The world as we perceive it is full of a rich variety: some of it is beautiful, some of it is ugly; parts seem to us good, parts bad. But all this has nothing to do with the purely causal properties of things, and it is the properties with which science is concerned. I am not suggesting that if we knew these properties completely we should have a complete knowledge of the world, for its concrete variety is an equally legitimate object of knowledge. What I am saying is that science is that sort of knowledge which gives causal understanding, and that this sort of knowledge can in all likelihood be completed, even where living bodies are concerned, without taking account of anything but their physical and chemical properties.

The Scientific Outlook
Chapter V (p. 133)

Santayana, George

Science, then is the alternative consideration of common experience; it is common knowledge extended and refined.

The Life of Reason
Part V, Chapter I (p. 393)

Shermer, Michael

Science is not the affirmation of a set of beliefs but a process of inquiry aimed at building a testable body of knowledge constantly open to rejection or confirmation. In science, knowledge is fluid and certainty fleeting. That is at the heart of its limitations. It is also its greatest strength.

Why People Believe Weird Things
Part 2
Pseudoscience and Superstition (p. 124)

Spencer, Herbert

Science is organized knowledge . . .

Education
Chapter II (p. 119)

Stewart, Ian

I may not understand it, but it sure looks important to me.

Does God Play Dice? (p. 121)

Sullivan, J.W.N.

Knowledge for the sake of knowledge, as the history of science proves, is an aim with an irresistible fascination for mankind, and which needs no defence. The mere fact that science does, to a great extent, gratify our intellectual curiosity, is a sufficient reason for its existence.

The Limitations of Science
Chapter 7, Section 3 (p. 164)

Szent-Györgyi, Albert

Knowledge is a sacred cow, and my problem will be how we can milk her while keeping clear of her horns.

Science
Teaching and the Expanding Knowledge (p. 1278)
Volume 146, Number 3649, December 4, 1964

Thomson, Sir George

Science is knowledge which, in principle at least, is public in the sense that it may be shared by many, unlike private personal experiences such as dreams or pain. This suggests a preference for statements which can be made in a form valid for large classes of possible observers.

The Inspiration of Science
Introduction (p. 6)

Trotter, William

Knowledge comes from noticing resemblances and recurrences in the events that happen around us.

In W.I.B. Beveridge
The Art of Scientific Investigation
Observation (p. 92)

Unknown

Right now I'm having amnesia and *déjà vu* at the same time. I think I've forgotten this before.

Source unknown

Waddington, C.H.

Scientific knowledge and understanding is a communal achievement, the sum of a multitude of contributions from many different people. Any individual may feel a certain justifiable pride if he knows that he has added one brick to the structure.

The Scientific Attitude
Science's Failure and Success (p. 62)

Weisskopf, Victor F.

Knowledge has to be sucked into the brain, not pushed into it.

The Privilege of Being a Physicist
Chapter 4 (p. 31)

Whitehead, Alfred North

Science is not discussing the causes of knowledge, but the coherence of knowledge.

The Concept of Nature
Teaching Science (p. 41)

The consequences of a plethora of half-digested theoretical knowledge are deplorable.

The Organization of Thought
The Aims of Education (p. 9)

Ziman, John

Scientific knowledge is not created solely by the piecemeal mining of discrete facts by uniformly accurate and reliable individual scientific investigations. The process of criticism and evaluation, of analysis and synthesis, are essential to the whole system. It is impossible for each one of us to be continually aware of all that is going on around us, so that we can immediately decide the significance of every new paper that is published. The job of making such judgments must therefore be delegated to the best and wisest among us, who speak, not with their own personal voices, but on behalf of the whole community of Science. Anarchy is as much a danger in that community as in any tribe or nation. It is impossible for the consensus—public knowledge—to be voiced at all, unless it is channelled through the minds of selected persons, and restated in their words for all to hear.

Public Knowledge
Chapter 7 (pp. 136–7)

LANGUAGE

Bloomfield, Leonard

The use of language in science is specialized and peculiar. In a brief speech the scientist manages to say things which in ordinary language would require a vast amount of talk. His hearers respond with great accuracy and uniformity. The range and exactitude of scientific prediction exceed any cleverness of everyday life: the scientist's use of language is strangely effective and powerful. Along with systematic observation, it is this peculiar use of language which distinguishes science from non-scientific behavior.

International Encyclopedia of Unified Science
Linguistic Aspects of Science
Volume 1, Number 4 (p. 1)

Hesse, Mary B.

. . . there is an external world which can in principle be exhaustively described in scientific language. The scientist, as both observer and language-user, can capture the external facts of the world in propositions that are true if they correspond to the facts and false if they do not. Science is ideally a linguistic system in which true propositions are in one-to-one relation to facts, including facts that are not directly observed because they involve hidden entities or properties, or past events or far distant events. These hidden events are described in theories, and theories can be inferred from observation, that is, the hidden explanatory mechanism of the world can be discovered from what is open to observation. Man as scientist is regarded as standing apart from the world and able to experiment and theorize about it objectively and dispassionately.

Revolutions and Reconstructions in the Philosophy of Science
Introduction (p. vii)

Huxley, Aldous

Like the man of letters, the scientist finds it necessary to "give a purer sense to the words of the tribe." But the purity of scientific language is not the same as the purity of literary language. The aim of the scientist is to say only one thing at a time, and to say it unambiguously and with the greatest possible clarity. To achieve this, he simplifies and jargonizes.

Literature and Science
Chapter 5 (p. 12)

Johnson, Samuel

I am not so lost in lexicography, as to forget that words are the daughters of earth, and that things are the sons of heaven. Language is only the instrument of science, and words are but the signs of ideas: I wish, however, that the instrument might be less apt to decay, and that signs might be permanent, like the things which they denote.

Dictionary of the English Language
Introduction

Kistiakowski, George B.

. . . science is today one of the few common languages of mankind; it can provide a basis for understanding and communication of ideas between people that is independent of political boundaries and ideologies.

Bulletin of the Atomic Scientists
Science and Foreign Affairs (p. 115)
Volume XVI, Number 4, April 1960

Wigner, Eugene

The simplicities of natural laws arise through the complexities of the languages we use for their expression.

Communications on Pure and Applied Mathematics
The Unreasonable Effectiveness of Mathematics
in the Natural Sciences (p. 1)
Volume 13, 1960

LAWS

Collingwood, R.G.

The scientist collects crude facts, but he stores only what he has converted them into: *laws*. Laws are the body of science. Laws are what it is a scientist's business to come at. Laws are what a master-scientist has to teach. Laws are what a pupil-scientist has to learn.

The New Leviathan
Part II, Chapter XXXI, aphorism 31.28

Einstein, Albert

A law cannot be definite for the one reason that the conceptions with which we formulate it develop and may prove insufficient in the future. There remains at the bottom of every thesis and of every proof some remainder of the dogma of infallibility.

Cosmic Religion
On Science (p. 100)

Friend, J.W.
Feibleman, James K.

If nature is not subject to law, then the whole of science is a fruitless proceeding.

What Science Really Means
Chapter IV (p. 95)

Hopper, Grace Murray

If you do something once, people will call it an accident. If you do it twice, they call it a coincidence. But do it a third time and you've just proven a natural law.

In Ethlie Ann Vare and Greg Ptacek
Mothers of Invention
From Eggbeaters to Eggheads (p. 187)

Gay-Lussac, Joseph Louis

If one were not animated with the desire to discover laws, they would often escape the most enlightened attention.

In Maurice Crosland
Gay-Lussac: Scientist and Bourgeois
Chapter 3 (p. 43)

Johnson, George

In *Zen and the Art of Motorcycle Maintenance*, Phaedrus, the author Robert Pirsig's alter ego, is sitting outside a motel room in the West, drinking whiskey with his traveling companions and listening to his son, Chris, tell ghost stories. "Do you believe in ghosts?" Chris asks his father. "No," Phaedrus says. "They contain no matter and have no energy and therefore, according to the laws of science, do not exist except in people's minds." Then he pauses and reflects: "Of course, the laws of science contain no matter and have no energy either and therefore do not exist except in people's minds."

Fire in the Mind
Phaedrus's Ghosts (p. 24)

Maxwell, James Clerk

The only laws of matter are those which our minds must fabricate, and the only laws of mind are fabricated for it by matter.

In Gerald Edelman
Bright Air, Brilliant Fire
Chapter 3 (p. 16)

Michelson, A.A.

The more important fundamental laws and facts of physical science have all been discovered, and these are now so firmly established that the possibility of their ever being supplanted in consequence of new discoveries is exceedingly remote . . . our future discoveries must be looked for in the sixth place of decimals.

Light Waves and Their Uses
Lecture II (pp. 23, 24)

Russell, Bertrand

The discovery of causal laws is the essence of science and therefore there can be no doubt that scientific men do right to look for them. If there is any region where there are no causal laws, that region is inaccessible to science. But the maxim that men of science should seek causal laws is as obvious as the maxim that mushroom-gatherers should seek mushrooms.

Religion and Science
Determinism (p. 153)

Thoreau, Henry David

If we knew all the laws of Nature, we should need only one fact, or the description of one actual phenomenon, to infer all the particular results at that point. Now we know only a few laws, and our result is vitiated, not, of course, by any confusion or irregularity in Nature, but by our ignorance of essential elements in the calculation. Our notions of law and harmony are commonly confined to those instances which we detect; but the harmony which results from a far greater number of seemingly conflicting, but really concurring, laws, which we have not detected, is still more wonderful. The particular laws are as our points of view, as, to the traveler, a mountain outline varies with every step, and it has an infinite number of profiles, though absolutely but one form. Even when cleft or bored through it is not comprehended in its entireness.

Walden
The Pond in Winter

No great discovery was ever made without a bold guess.
Sir Isaac Newton – (See p. 38)

LITERATURE

Bulwer-Lytton, Edward

In science, read, by preference, the newest works; in literature, the oldest. The classic literature is always modern.

Caxtoniana
Hints on Mental Culture (p. 110)

In science, address the few; in literature the many. In science, the few must dictate opinion to the many; in literature, the many, sooner or later, force their judgment on the few.

Caxtoniana
Readers and Writers (p. 428)

Carson, Rachel

The aim of science is to discover and illuminate truth. And that, I take it, is the aim of literature, whether biography or history or fiction; it seems to me, then, that there can be no separate literature of science.

In Paul Brooks
The House of Life: Rachel Carson at Work
Fame (p. 128)

Compton, Karl Taylor

I think the story of research development is well described by the ditty:

Little drops of water,
Little grains of sand,
Make the mighty ocean
And the pleasant land.

Anyone familiar with scientific literature realizes the enormous number of contributions, most of them small and not very significant, but each and all gradually raising the level of understanding in the storehouse of knowledge until finally the stage is reached at which a great scientific discovery or a mighty practical application can be made.

A Scientist Speaks (p. 62)

Crothers, Samuel McChord

The distinction between Literature and Science is fundamental. What is a virtue in one sphere is a vice in the other.

The Gentle Reader
The Hinter-Land of Science (p. 229)

Dickinson, G. Lowes

When Science arrives, it expels Literature.

In S. Chandrasekhar
Truth and Beauty
Chapter Three (p. 55)

Huxley, Thomas

Science and literature are not two things, but two sides of one thing.

In Leonard Huxley
Life and Letters Of Thomas Henry Huxley
Volume I
Chapter XVI (p. 231)

James, William

The 'marvels' of Science, about which so much edifying popular literature is written, are apt to be 'caviare' to the men in the laboratories.

Principles of Psychology
Volume 2
The Emotions
No Special Brain-Centres for Emotion (p. 472)

Levine, George

Once one is committed to the view that science is not so clearly separable from the human sciences . . . or from other humanist enterprises, history of science begins to blur with social history. Literature becomes part of the history of science. Science is reflected in literature. And the tools of literary criticism become instruments in the understanding of scientific discourse.

One Culture: Essays in Science and Literature
One Culture: Science and Literature, II (p. 22)

Overhage, Carl F.J.

The public printed record of the results of scholarly research is the universal device that transcends the barriers of space and time between scholars. It makes the most recent advances of human knowledge accessible to students and scholars throughout the world. Wherever there is a library, any person who has learned the language may participate in the outstanding intellectual adventures of his time. The same record extends into the past; through an unbroken sequence of communications, the scholars of today can trace the origin of a new concept in different periods and in different countries. By standing on the shoulders of a giant, he may see farther.

The wide availability of the record is one of the guarantees of its soundness. In science especially, truth is held to reside in findings that can be experimentally verified anywhere, at any time.

Science
Science Libraries: Prospects and Problems (p. 804)
Volume 155, Number 3764, February 1967

President's Science Advisory Committee
Science and technology can flourish only if each scientist interacts with his colleagues and his predecessors, and only if every branch of science interacts with other branches of science; in this sense science must remain unified if it is to remain effective. The ideas and data that are the substance of science and technology are embodied in the literature; only if the literature remains a unity can science itself be unified and viable.

Science, Government and Information
Part 1 (p. 7)

Valéry, Paul
"Science" means simply the aggregate of all the recipes that are *always successful*. All the rest is . . . literature.

The Collected Works of Paul Valéry
Volume 14
Analects (p. 64)

Weinberg, Alvin
. . . the scientific community has evolved an empirical method for establishing scientific priorities—for deciding what is important in science and what is unimportant. This is the scientific literature. The process of self-criticism, which is integral to the literature of science, is one of the most characteristic features of science. Nonsense is weeded out and held up to ridicule in the literature, whereas what is worthwhile receives much sympathetic attention. This process of self-criticism embodied in the literature, though implicit is nonetheless real and highly significant. The existence of a healthy, viable, refereed scientific literature in itself helps assure society that the science it supports is valid and deserving of support. This is a most important, though little recognized, social function of the scientific literature.

Reflections on Big Science
Chapter III (p. 70)

LOGIC

Jowett, Benjamin
Logic is neither a science nor an art, but a dodge.

In James R. Newman
The World of Mathematics
Volume Four (p. 2402)

Schiller, F.C.S.
Among the obstacles to scientific progress a high place must certainly be assigned to the analysis of scientific procedure which Logic has provided. . . . It has not tried to describe the methods by which the sciences have actually advanced, and to extract . . . rules which might be used to regulate scientific progress, but . . . has freely re-arranged the actual procedure in accordance with its prejudices. For the order of discovery there has been substituted an order of 'proof' . . .

In Charles Singer (Editor)
Studies in the History and Method of Science
Volume I
Scientific Discovery and Logical Proof (p. 235)

. . . it is not too much to say that the more deference men of science have paid to Logic, the worse it has been for the scientific value of their reasoning. . . . Fortunately for the world, however, the great men of science have usually been kept in salutary ignorance of the logical tradition . . .

In Charles Singer (Editor)
Studies in the History and Method of Science
Volume I
Scientific Discovery and Logical Proof (p. 236)

Selye, Hans
. . . logic is to Nature as a guide is to a zoo. The guide knows exactly where to locate the African lion, the Indian elephant or the Australian kangaroo, once they have been captured, brought together and labeled

for inspection. But this kind of knowledge would be valueless to the hunter who seeks them in their natural habitat. Similarly, logic is not the key to Nature's order but only the catalogue of the picture gallery in man's brain where his impressions of natural phenomena are stored.

From Dream to Discovery
How to Think (p. 266)

Unknown
Reiteration of an argument is often more effective than its inherent logic.

Source unknown

Whitehead, Alfred North
Neither logic without observation, nor observation without logic, can move one step in the formation of science.

The Organisation of Thought
Chapter VI (p. 132)

Logic is the olive branch from the old to the young, the wand which in the hands of youth has the magic property of creating science.

The Organisation of Thought
Chapter VI (p. 133)

MAGIC

Asimov, Isaac

Once upon a time, there were priesthoods of magic, and members of those priesthoods cast spells, muttered runes, and made intricate diagrams on the floor with powders of arcane composition . . . Nowadays, there is a modern priesthood of science that calls on the power of expanding steam, of shifting electrons or drifting neutrons, of exploding gasoline or uranium, and does so without spells, powders, or even any visible change of expression. In response, onlookers are without awe, for indeed, they seem to participate in the magic.

New York Times Book Review
November 19, 1967
Review of the book *The Way Things Work*, Volume 1

Bronowski, Jacob

Man masters nature not by force but by understanding. That is why science has succeeded where magic failed because it has looked for no spell to cast on nature.

Science and Human Values
The Creative Mind (p. 18)

Chesterton, G.K.

All the terms in the science books, 'law', 'necessity', 'order', 'tendency', and so on, are really unintellectual, because they assume an inner synthesis which we do not possess. The only words that ever satisfied me as describing Nature are the terms used in fairy books, 'charm', 'spell', 'enhancement'. A tree grows because it is a magic tree. Water runs down because it is bewitched. The sun shines because it is bewitched.

Orthodoxy
The Ethics of Elfland (p. 94)

Cooper, Leon

A theory is a well-defined structure hopefully in correspondence with what we observe . . . It's an architecture, a cathedral. There is an ancient human longing to impose rational order on a chaotic world. The detective does it, the magician does it. That's why people love Sherlock Holmes. Science came out of magic. Science is the modern expression of what the ancient magician did. The world is a mess, and people want it to be orderly.

In George Johnson
In the Palaces of Memory
The Memory Machine (pp. 114–5)

Farb, Peter
In place of science, the Eskimo has only magic to bridge the gap between what he can understand and what is not known. Without magic, his life would be one long panic.

Man's Rise to Civilization
Chapter III (p. 48)

Goodwin, Brian
There is no truth beyond magic . . . reality is strange. Many people think reality is prosaic. I don't. We don't explain things away in science. We get closer to the mystery.

In Roger Lewin
Complexity: Life at the Edge of Chaos
Chapter Two (p. 32)

Gould, Laurence M.
Science is not a form of black magic. A thousand blind alleys must often be explored before a right road is found; a thousand amateurs must have their fling before a Darwin or an Einstein comes along.

UNESCO Courier
Science and the Culture of Our Times (p. 4)
February 1968

Hsu, Francis L.K.
. . . to achieve popular acceptance, magic has to be dressed like science in America, while science has to be cloaked as magic in Hsi-ch'eng.

Health, Culture and Community
Part 2
A Cholera Epidemic in a Chinese Town (p. 149)

Rapoport, Anatol
Magic is essentially metaphorical. So are dreams. So is most artistic activity. Finally, theoretical science is essentially disciplined exploitation of metaphor.

Operational Philosophy
Chapter 17 (p. 203)

Watson, David Lindsay

"Scientific method" is, all too often, degenerating into a magic which employs the empty rituals of mathematics and measurement and the tabus of a worthless "logical rigour."

Scientists Are Human
Introduction (p. xiv)

Applied Science is a conjurer, whose bottomless hat yields impartially the softest of Angora rabbits and the most petrifying of Medusas.
Aldous Huxley – (See p. 15)

MAN OF SCIENCE

Bradley, Omar
With the monstrous weapons man already has, humanity is in danger of being trapped in this world by its moral adolescents. Our knowledge of science has already outstripped our capacity to control it. We have many men of science, too few men of God.

Address in Boston
November 10, 1948

Butler, Samuel
[Science] If it tends to thicken the crust of ice on which, as it were, we are skating, it is all right. If it tries to find, or professes to have found, the solid ground at the bottom of the water it is all wrong.

I do not know whether my distrust of men of science is congenital or acquired, but I think I should have transmitted it to descendants.

In Geoffrey Keynes and Brian Hill (Editors)
Samuel Butler's Notebooks
Science (p. 110)

[Men of science] If they are worthy of the name, are indeed about God's path and about his bed and spy out all his ways.

In Geoffrey Keynes and Brian Hill (Editors)
Samuel Butler's Notebooks
Men of Science (p. 204)

Chesterton, G.K.
Far away in some strange constellation in skies infinitely remote, there is a small star, which astronomers may some day discover. At least, I could never observe in the faces or demeanour of most astronomers or men of science any evidence that they had discovered it; though as a matter of fact they were walking about on it all the time. It is a star that brings forth out of itself very strange plants and very strange animals; and none stranger than the men of science.

The Everlasting Man
Chapter I (p. 23)

Darwin, Charles

Children are one's greatest happiness, but often and often a still greater misery. A man of science ought to have none—perhaps not a wife; for then there would be nothing in this wide world worth caring for, and a man might (whether he could is another question) work away like a Trojan.

More Letters of Charles Darwin
Volume I
Letter to J.L.A. de Quatrefages
July 11, 1862 (p. 202)

du Noüy, Pierre Lecomte

The man of science who cannot formulate a hypothesis is only an accountant of phenomena.

The Road to Reason
Chapter 3 (p. 77)

Einstein, Albert

. . . the man of science is a poor philosopher.

Out of My Later Years
Physics and Reality
I (p. 59)

Huxley, Thomas

The man of science has learned to believe in justification, not by faith, but by verification.

Collected Essays
Volume I
Method and Results
On Improving Natural Knowledge (p. 41)

The man of science knows that here, as everywhere, perfect order is manifested; that not a curve of the waves, not a note in the howling chorus, not a rainbow glint on a bubble which is other than a necessary consequence of the ascertained laws of nature; and that with sufficient knowledge of the conditions competent physicomathematical skill could account for, and indeed predict every one of those "chance" events.

In F. Darwin
The Life and Letters of Charles Darwin
Volume I (pp. 553–5)

Langley, John Newport

Those who have occasion to enter into the depths of what is oddly, if generously, called the literature of a scientific subject, alone know the difficulty of emerging with an unsoured disposition. The multitudinous facts presented by each corner of Nature form in large part the scientific man's burden to-day, and restrict him more and more, willy-nilly, to a narrower and narrower specialism. But that is not the whole of his burden. Much that he is forced to read consists of records of defective experiments, confused statements of results, wearisome description of detail, and unnecessarily protracted discussion of unnecessary hypotheses. The publication of such matter is a serious injury to the man of science; it absorbs the scanty funds of his libraries, and steals away his poor hours of leisure.

Report of the British Association for the Advancement of Science
1899
Presidential Address to the Physiology Section

Laplace, Pierre
The isolated man of science can dedicate himself without fear to dogmatism; he hears only from afar contradictions of his ideas. But in a scientific society the impact of dogmatic ideas soon results in their destruction, and the desire to win one another over to their point of view establishes necessarily among members the convention of admitting only the results of observations and calculation.

In Maurice Crosland
Gay-Lussac: Scientist and Bourgeois
Chapter 2 (p. 34)

Melville, Herman
. . . a man of true science . . . uses but few hard words, and those only when none other will answer his purpose; whereas the smatterer in science . . . thinks, that by mouthing hard words, he proves that he understands hard things.

White Jacket
Chapter LXIII (p. 277)

More, Louis Trenchard
. . . so long as men of science restrict their endeavor to the world of material substance and material force, they will find that their field is practically without limits, so vast and so numerous are the problems to be solved. And it should distress no one to discover that there are other fields of knowledge in which science is not concerned; on the contrary, the fact that the range of science is limited should encourage us to greater hopes, because our freedom of action is still far greater than our powers of accomplishment. After centuries of effort, the ocean of the unknown lies before us unexplored.

The Limitations of Science
Chapter VII (p. 261)

Oppenheimer, J. Robert

Both the man of science and the man of art live always at the edge of mystery, surrounded by it. Both, as the measure of their creation, have always had to do with the harmonization of what is new with what is familiar, with the balance between novelty and synthesis, with the struggle to make partial order in total chaos . . . This cannot be an easy life.

In Robert Jungk
Brighter than a Thousand Suns
Chapter Twenty (pp. 333–4)

Poincaré, Henri

The true man of science has no such expression in his vocabulary as useful science . . . if there can be no science for science's sake there can be no science.

Technology and Culture
In James Kip Finch
Engineering and Science (p. 330)
Fall 1961

Popper, Karl R.

It is not his *possession* of knowledge, of irrefutable truth, that makes the man of science, but his persistent and recklessly critical *quest* for truth.

The Logic of Scientific Discovery
Chapter 10, Section 85 (p. 281)

Renan, Ernest

With the saints, the heroes, the great men of all ages we may fearlessly compare our men of scientific minds, given solely to the research of truth, indifferent to fortune, often proud of their poverty, smiling at the honors they are offered, as careless of flattery as of obloquy, sure of the worth of that they are doing, and happy because they possess truth.

Scientific American
The Nobility of Science
Volume XL, Number 20, New Series (p. 310)
May 17, 1879

Ross, Sir Ronald

A witty friend of mine once remarked that the world thinks of the man of science as one who pulls out his watch and exclaims, "Ha! half an hour to spare before dinner: I will just step down to my laboratory and make a discovery." Who but men of science themselves are to blame for such a misconception? Out of the many memoirs which fill our libraries few recount the labours of investigators, even of those who seek to solve the secrets of the great maladies which annually destroy millions of us—surely a matter of interest to everyone. Our books of science are records of results rather than of that sacred passion for discovery which leads to them. Yet many discoveries have really been the climax of an intense drama, full of hopes and despairs, visions seen in darkness, many failures, and a final triumph: in which the protagonists are man and nature, and the issue a decision for all the ages.

Memoirs
Preface (pp. v–vi)

Russell, Bertrand
The man of science looks for facts that are significant, in the sense of leading to general laws; and such facts are frequently quite devoid of intrinsic interest.

The Scientific Outlook
Chapter I (p. 49)

All the conditions of happiness are realized in the life of the man of science.

The Conquest of Happiness
X (p. 146)

Spencer, Herbert
Only the sincere man of science (and by this title we do not mean the mere calculator of distances, or analyser of compounds, or labeller of species; but him who through lower truths seeks higher, and eventually the highest)—only the genuine man of science, we say, can truly know how utterly beyond, not only human knowledge, but human conception, is the Universal Power of which Nature, and Life, and Thought are manifestations.

Education
Chapter I (p. 85)

Suits, C.G.
I've never met that "coldly calculating man of science" whom the novelists extol . . . I doubt that he exists; and if he did exist I greatly fear that he would never make a startling discovery or invention.

In Frederic Brownell
The American Magazine
Heed That Hunch (p. 142)
December 1945

Sullivan, J.W.N.
. . . outside their views on purely scientific matters there is nothing *characteristic* of men of science.

Aspects of Science
Scientific Citizen (p. 120)

Whitehead, Alfred North

No man of science wants merely to know. He acquires knowledge to appease his passion for discovery. He does not discover in order to know, he knows in order to discover.

The Organisation of Thought
Chapter II (p. 37)

Whitney, W.R.

For the engineer "safety first" is a good slogan, but "safety last" is better for the man of research.

Science
The Stimulation of Research in Pure Science which has Resulted from the Needs of Engineers and of Industry (p. 289)
Volume LXV, Number 1862, March 25, 1927

A scientist is a person who knows more and more about less and less, until he knows everything about nothing.
John Ziman – (See p. 329)

MEASUREMENT

Asimov, Isaac

. . . we must remember that measures were made for man and not man for measures.

Of Time and Space and Other Things
Forget It (p. 143)

Balfour, A.J.

Science depends upon measurement, and things not measurable are therefore excluded, or tend to be excluded, from its attention.

Address, 1917
National Physical Laboratory
In William H. George
The Scientist in Action
Some Problems in Theorizing (pp. 263–4)

Bondi, Hermann

A quantity like time, or any other physical measurement, does not exist in a completely abstract way. We find no sense in talking about something unless we specify how we measure it. It is the definition by the method of measuring a quantity that is the one sure way of avoiding talking nonsense . . .

Relativity and Common Sense
Chapter VII (p. 65)

Brouwer, L.E.J.

It is well to notice in this connection that a natural law in the statement of which measurable magnitudes occur can only be understood to hold in nature with a certain degree of approximation; indeed natural laws as a rule are not proof against sufficient refinement of the measuring tools.

Bulletin of the American Mathematical Society
Intuitionism and Formalism (p. 82)
Volume 20, November 1913

Deming, William Edwards
It is important to realize that it is not the one measurement, alone, but its relation to the rest of the sequence that is of interest.

Statistical Adjustment of Data
Chapter I (p. 3)

Dewey, John
Insistence upon numerical measurement, when it is not inherently required by the consequence to be effected, is a mark of respect for the ritual of scientific practice at the expense of its substance.

Logic: The Theory of Inquiry
Quantity and Measure (p. 205)

Fox, Russell
Gorbuny, Max
Hooke, Robert
. . . measurement is science's highest court of appeal, pronouncing its final verdict for or against the meekest and the loftiest ideas alike.

The Science of Science
Chapter 3 (p. 20)

Isaiah 40:12
Who was it measured the waters of the sea in the hollow of his hand and calculated the dimensions of the heavens, gauged the whole earth to the bushel, weighed the mountains in scales, the hills in a balance?

The Bible

Kaplan, Abraham
Measurement, we have seen, always has an element of error in it. The most exact description or prediction that a scientist can make is still only approximate. If, as sometimes happens, a perfect correspondence with observation does appear, it must be regarded as accidental, and, as Jevons [see *The Principles of Science*, p. 457] . . . remarks, it "should give rise to suspicion rather than to satisfaction."

The Conduct of Inquiry
Chapter VI, Section 25 (p. 215)

Proleptically, I would say that whether we can measure something depends, not on that thing, but on how we have conceptualized it, on our knowledge of it, above all on the skill and ingenuity which we can bring to bear on the process of measurement which our inquiry can put to use.

The Conduct of Inquiry
Chapter V, Section 20 (p. 176)

Lewis, Gilbert N.

I have no patience with attempts to identify science with measurement, which is but one of its tools, or with any definition of the scientist which would exclude a Darwin, a Pasteur, or a Kekulé.

The Anatomy of Science (p. 6)

Pietschmann, H.

The model that science constructs appears to us more serious, more important and more real than experienced reality. Contradictions are not merely eliminated; they become mistakes that are regarded as a breakdown. Thus today we must expand Galileo's description of modern scientific method by saying: to measure everything measurable; to make everything which is not measurable measurable; and to deny everything which cannot be made measurable.

Das Ende des naturwissenschaftlichen Zeitalters (p. 29)

Pindar

But in everything is there due measure.

The Extant Odes of Pindar
Olympia 13 (pp. 45–6)

Reynolds, H.T.

Crude measurement usually yields misleading, even erroneous conclusions no matter how sophisticated a technique is used.

Analysis of Nominal Data (p. 56)

METAPHOR

Fulford, Robert
Metaphor, the life of language, can be the death of meaning. It should be used in moderation, like vodka. Writers drunk on metaphor can forget they are conveying information and ideas.

Globe & Mail (Toronto)
December 4, 1996

Harré, Rom
Metaphor and simile are the characteristic tropes of scientific thought, not formal validity of argument.

Varieties of Realism
Part I
Locating Realism (p. 7)

METHOD

Born, Max
There are two objectionable types of believers: those who believe the incredible and those who believe that 'belief' must be discarded and replaced by 'the scientific method.'

Natural Philosophy of Cause and Chance
Appendix One (p. 209)

Camus, Albert
When one has no character one *has* to apply a method.

The Fall (p. 11)

Cohen, Morris R.
Nagel, Ernest
. . . the safety of science depends on there being men who care more for the justice of their methods than for any results obtained by their use.

An Introduction to Logic and Scientific Method
Chapter XX, Section 2 (p. 402)

Committee on the Conduct of Science, National Academy of Sciences
The fallibility of methods means that there is no cookbook approach to doing science, no formula that can be applied or machine that can be built to generate scientific knowledge . . . The skillful application of methods to a challenging problem is one of the great pleasures of science.

On Being a Scientist (p. 6)

Some methods, such as those governing the design of experiments or the statistical treatment of data, can be written down and studied. But many methods are learned only through personal experience and interactions with other scientists. Some are even harder to describe or teach. Many of the intangible influences on scientific discovery—curiosity, intuition, creativity—largely defy rational analysis, yet they are often the tools that scientists bring to their work.

On Being a Scientist (p. 6)

Doyle, Sir Arthur Conan

Pon my word Watson, you are coming along wonderfully. You have really done very well indeed. It is true that you have missed everything of importance, but you have hit upon the method . . .

The Complete Sherlock Holmes
A Case of Identity

Egler, Frank E.

A science is to a great extent the product of the methods applied to it. We find only what we look for. Wc strain out only what our strainer is designed to detain. We net only those fish neither too small to escape, nor too large but that they break the net. We photograph only what passes thru the lens and affects the film. We measure only that what is measurable. Indeed, science is but an artifact of its methodology. Nothing more.

The Way of Science (p. 35)

Gay-Lussac, Joseph Louis

We are convinced that exactitude in experiments is less the outcome of faithful observation of the divisions of an instrument than of the exactitude of method.

In Maurice Crosland
Gay-Lussac: Scientist and Bourgeois
Chapter 3 (p. 70)

Goethe, Johann Wolfgang von

Content without method leads to fantasy; method without content to empty sophistry; matter without form to unwieldy crudition, form without matter to hollow speculation.

Scientific Studies
Volume 12
Chapter VIII (p. 306)

Hilbert, David

He who seeks for methods without having a definite problem in mind seeks for the most part in vain.

Bulletin of the American Mathematical Society
Hilbert: Mathematical Problems (p. 444)
Volume 8

Hubble, Edwin

The methods of science may be described as the discovery of laws, the explanation of laws by theories, and the testing of theories by new observations. A good analogy is that of the jigsaw puzzle, for which the laws are the individual pieces, the theories local patterns suggested by a few pieces, and the tests the completion of these patterns with pieces previously unconsidered.

The Nature of Science
The Nature of Science (p. 11)

Huxley, Thomas
No delusion is greater than the notion that method and industry can make up for lack of motherwit, either in science or in practical life.

Collected Essays
Volume I
Method and Results
The Progress of Science (p. 46)

I am not afraid of the priests in the long-run. Scientific method is the white ant which will slowly but surely destroy their fortifications. And the importance of scientific methods in modern practical life—always growing and increasing—is the guarantee for the gradual emancipation of the ignorant upper and lower classes, the former of whom especially are the strength of the priests.

Collected Essays
Volume III
Life and Letters (p. 330)

Krutch, Joseph Wood
We are committed to the scientific method and measurement is the foundation of that method; hence we are prone to assume that whatever is measurable must be significant and that whatever cannot be measured may as well be disregarded.

Human Nature and the Human Condition
Chapter 5 (p. 78)

Medawar, Sir Peter
"The scientific method," as it is sometimes called, is a potentiation of common sense.

Advice to a Young Scientist
Chapter 11 (p. 93)

Pearson, Karl
I assert that the encouragement of scientific investigation and the spread of scientific knowledge by largely inculcating scientific habits of mind will lead to more efficient citizenship and so to increased social stability. Minds trained to scientific methods are less likely to be led by mere appeal to the passions or by blind emotional excitement to sanction acts which in the end may lead to social disaster.

The Grammar of Science
Introductory
Section 3 (p. 13)

Now this is the peculiarity of scientific method, that when once it has become a habit of mind, that mind converts *all* facts whatsoever into science. The field of science is unlimited; its solid contents are endless, every group of natural phenomena, every phase of social life, every stage of past or present development is material for science. *The unity of all science consists alone in its method, not in its material*. The man who classifies facts of any kind whatever, who sees their mutual relation and describes their sequence, is applying the scientific method and is a man of science. The facts may belong to the past history of mankind, to the social statistics of our great cities, to the atmosphere of the most distant stars, to the digestive organs of a worm, or to the life of a scarcely visible bacillus. It is not the facts themselves which form science, but the method in which they are dealt with. The material of science is co-extensive with the whole physical universe, not only that universe as it now exists, but with its past history and the past history of all life therein. When every fact, every present or past phenomenon of that universe, every phase of present or past life therein, has been examined, classified, and co-ordinated with the rest, then the mission of science will be completed. What is this but saying that the task of science can never end till man ceases to be, till history is no longer made, and development itself ceases.

The Grammar of Science
Introductory
Section 5 (p. 16)

Pirsig, Robert M.

Traditional scientific method has always been at the very *best*, 20–20 hindsight. It's good for seeing where you've been.

Zen and the Art of Motorcycle Maintenance
Part III, Chapter xxiv (p. 273)

Pólya, George

My method to overcome a difficulty is to go round it.

How to Solve It (p. 181)

Russell, Bertrand

Whatever knowledge is attainable must be attainable by scientific methods; and what science cannot discover, mankind cannot know.

Religion and Science
Science and Ethics (p. 243)

Scientific method . . . consists in observing such facts as will enable the observer to discover general laws governing facts of the kind in question.

The Scientific Outlook
Chapter I (p. 15)

Scientific method . . . consists mainly in eliminating those beliefs which there is reason to think a source of shocks, while retaining those against which no definite argument can be brought.

Human Knowledge: Its Scope and Limits
Chapter III (p. 201)

Tennant, F.R.

Half a century ago, it was taught that the scientific method is the sole means of approach to the whole realm of possible knowledge: that there were no reasonably propounded questions worth discussing to which its method was inapplicable. Such belief is less widely held today. Since many men of science became their own epistemologists, science has been more modest.

Philosophical Theology
Volume I
Chapter XIII (p. 333)

Thomson, Sir George

The scientific method is not a royal road leading to discoveries in research, as Bacon thought, but rather a collection of pieces of advice, some general some rather special, which may help to guide the explorer in his passage through the jungle of apparently arbitrary facts.

The Inspiration of Science
The Scientific Method (p. 7)

Wilson, Edwin B.

A method is a dangerous thing unless its underlying philosophy is understood, and none more dangerous than the statistical. Our aim should be, with care, to avoid in the main erroneous conclusions. In a mathematical and strictly logical discipline the care is one of technique; but in the natural sciences and in statistics the care must extend not only over the technique but to the matter of judgment, as is necessarily the case in coming to conclusions upon any problem of real life where the complications are great. Over-attention to technique may actually blind one to the dangers that lurk about on every side—like the gambler who ruins himself with his system carefully elaborated to beat the game. In the long run it is only clear thinking, experienced methods, that win the strongholds of science.

Science
The Statistical Significance of Experimental Data (p. 94)
Volume 58, Number 1493, August 1923

MYTH

Bernal, J.D.
Science, on the one hand, is ordered technique; on the other, it is rationalized mythology.

Science in History
Preface (p. ix)

Hubbard, Ruth
The mythology of science asserts that with many different scientists all asking their own questions and evaluating the answers independently, whatever personal bias creeps into their individual answers is canceled out when the large picture is put together. This might conceivably be so if scientists were women and men from all sorts of different cultural and social backgrounds who came to science with very different ideologies and interests. But since, in fact, they have been predominantly university-trained white males from privileged social backgrounds, the bias has been narrow and the product often reveals more about the investigator than about the subject being researched.

Women Look at Biology Looking At Women
Have Only Men Evolved? (p. 31)

Mahadeva, M.
Myths are errors that result both from scientists bringing societal preconceptions into science and from scientists feeding society ideas that masquerade as science.

Science Teacher
From Misinterpretations to Myths
Volume 56, Number 4, 1989

Popper, Karl R.
Thus science must begin with myths—and with the criticism of myths . . .

In C.A. Mace (Editor)
British Philosophy in the Mid-Century
Philosophy of Science: A Personal Report
VII (p. 177)

Science never starts from scratch; it can never be described as free from assumptions; for at every instant it presupposes a horizon of expectations—yesterday's horizon of expectations, as it were. Today's science is built upon yesterday's science [and so it is the result of yesterday's searchlight]; and yesterday's science, in turn, is based on the science of the day before. And the oldest scientific theories are built on pre-scientific myths, and these, in their turn, on still older expectations.

Objective Knowledge: An Evolutionary Approach
Appendix (pp. 346–7)

Whyte, Lancelot Law
In the ultimate analysis science is born of myth and religion, all three being expressions of the ordering spirit of the human mind.

The Unconscious Before Freud (pp. 82–3)

. . . But do [something] a third time and you've just proven a natural law.
Grace Murray Hopper – (See p. 110)

OBSERVATION

Arp, Halton

Of course, if one ignores contradictory observations, one can claim to have an "elegant" or "robust" theory. But it isn't science.

Science News
Letters (p. 51)
Volume 140, Number 4, July 27, 1991

da Vinci, Leonardo

Science is the observation of things possible, whether present or past; prescience is the knowledge of things which may come to pass.

The Literary Works of Leonardo da Vinci
Volume II, 1148 (p. 239)

Eddington, Sir Arthur

For the truth of the conclusions of physical science, observation is the supreme Court of Appeal. It does not follow that every item which we confidently accept as physical knowledge has actually been certified by the Court; our confidence is that it would be certified by the Court if it were submitted. But it does follow that every item of physical knowledge is of a form which might be submitted to the Court. It must be such that we can specify (although it may be impracticable to carry out) an observational procedure which would decide whether it is true or not. Clearly a statement cannot be tested by observation unless it is an assertion about the results of observation. Every item of physical knowledge must therefore be an assertion of what has been or would be the result of carrying out a specified observational procedure.

The Philosophy of Physical Science (pp. 9–10)

Fourier, J.B.

The primordial causes are unknown to us; but they are subject to simple and constant laws which one can discover through observation, and whose study is the object of natural philosophy.

In W. Thomson and P.G. Tait
Treatise on Natural Philosophy
Part I
Preface (p. v)

Gregg, Alan

. . . most of the knowledge and much of the genius of the research worker lie behind selection of what is worth observing. It is a crucial choice, often determining the success or failure of months of work, often differentiating the brilliant discoverer from the . . . plodder.

The Furtherance of Medical Research
Medical Research Described (p. 8)

Herschel, John Frederick William

There is scarcely any well-informed person, who, if he has but the will, has not also the power to add something essential to the general stock of knowledge, if he will only observe regularly and methodically some particular class of facts which may most excite his attention . . .

A Preliminary Discourse on the Study of Natural Philosophy
Chapter IV, Section 128 (p. 133)

Jeffreys, H.

An observation, strictly, is only a sensation. Nobody means that we should reject everything but sensations. But as soon as we go beyond sensations we are making inferences.

Theory of Probability
General Questions (p. 412)

Moulton, Lord

When we are reduced to observation Science crawls.

In Alan Gregg
The Furtherance of Medical Research
Medical Research Described (p. 7)

Plantamour

Try to verify any law of nature and you will find that the more precise your observations, the more certain they will be to show irregular departure from the law.

Recherches expérimentales sur le mouvement simultané d'un pendule et de ses supports (pp. 3–4)

Spencer, Herbert
Every science begins by accumulating observations, and presently generalizes these empirically; but only when it reaches the stage at which its empirical generalizations are included in a rational generalization does it become developed science.

The Data of Ethics
Chapter IV (p. 71)

It is easier for a man to get funds for what he proposes to do than for what he is doing.
Knight Dunlap – (See p. 70)

OCCAM'S RAZOR

Aristotle

When the consequences of either assumption are the same, we should always assume that things are finite rather than an infinite number.

Physics
VIII vi 259a9–12

Dante Alighieri

. . . everything superfluous is unpleasing to God and to Nature.

De Monarchia
Cap. I, 14

Dixon, Malcom

God doesn't always shave with Occam's razor.

Attributed
In David Hall
New Scientist
Letters
God's Razor (p. 51)
Volume 142, Number 1922, April 23, 1994

Jeans, Sir James

When two hypotheses are possible, we provisionally choose that which our minds adjudge to be the simpler, on the supposition that this is more likely to lead in the direction of the truth. It includes as a special case the principle of Occam's razor—*Entia non multiplicanda praeter necessitatem*.

Physics and Philosophy
Chapter VII (p. 183)

Newton, Sir Isaac

We are to admit no more causes of natural things than such as are both true and sufficient to explain their appearances.

The Mathematical Principles of Natural Philosophy
Book Three
Rule I (p. 270)

Unknown

Simples sigillum veri
Cut causes, be merry
Slash 'em and dock 'em
Said William of Ockham
Wiping his razor
On the sleeve of his blazer

Times Literary Supplement
June 18, 1981 (p. 688)

Knowledge has to be sucked into the brain, not pushed into it.
Victor F. Weisskopf – (See p. 107)

PARADIGM

Barnes, Barry

. . . paradigms, the core of the culture of science, are transmitted and sustained just as is culture generally: scientists accept them and become committed to them as a result of training and socialization, and the commitment is maintained by a developed system of social control.

In Quentin Skinner (Editor)
The Return of Grand Theory in the Human Sciences
Thomas Kuhn (p. 89)

Kuhn, Thomas S.

The operations and measurements that a scientist undertakes in the laboratory are not "the given" of experience but rather "the collected with difficulty." They are not what the scientist sees—at least not before his research is well advanced and his attention focused . . . Science does not deal in all possible laboratory manipulations. Instead, it selects those relevant to the juxtaposition of a paradigm with the immediate experience that that paradigm has partially determined.

The Structure of Scientific Revolutions
Chapter X (p. 126)

A paradigm is what members of the scientific community share, *and*, conversely a scientific community consists of men who share a paradigm.

The Structure of Scientific Revolutions
Postscript (p. 176)

PARADOX

Cudmore, L.L. Larison
It is a bizarre paradox we are facing, for we find that experimental scientists (who are supposed to be fair) at times make the Spanish Inquisition a model of fair hearings and unbiased judgment.

The Center of Life
Cellular Evolution (p. 55)

Falletta, Nicholas
A paradox is truth standing on its head to attract attention.

The Paradoxicon (p. xvii)

Kelvin, William Thomson, Baron
In science there are no paradoxes.

In S.P. Thompson
The Life of William Thomson, Baron Kelvin of Largs (p. 833)

Rapoport, Anatol
Paradoxes have played a dramatic part in intellectual history, often foreshadowing revolutionary developments in science, mathematics, and logic. Whenever, in any discipline, we discover a problem that cannot be solved within the conceptual framework that supposedly should apply, we experience an intellectual shock. The shock may compel us to discard the old framework and adopt a new one. It is to this process of intellectual molting that we owe the birth of many of the major ideas in mathematics and science.

Scientific American
Escape from Paradox (p. 50)
Volume 217, Number 1, July 1967

Russell, Bertrand
Although this may seem a paradox; all exact science is dominated by the idea of approximation.

In Jefferson Hane Weaver
The World of Physics
Volume II (p. 22)

PRAYER

Lewis, Sinclair
God give me unclouded eyes and freedom from haste. God give me a quiet and relentless anger against all pretense and all pretentious work and all work left slack and unfinished. God give me a restlessness whereby I may neither sleep nor accept praise till my observed results equal my calculated results or in pious glee I discover and assault my error. God give me the strength not to trust to God!

Arrowsmith
Chapter XXVI, Section II (p. 280)

Selye, Hans
Almighty Drive who, through the ages,
Has kept men trying to master Nature by understanding,
Give me faith—for that is what I need most now

This is a rare and solemn moment in my life:
I stumbled across what seems to be
A new path into the unknown.
A road that promises to lead me closer to You:
The law behind the unknown

. . .

I cannot know whether You listen,
But I do know that I must pray.

From Dream to Discovery
Who Should Do Research (p. 41)

PREDICTION

Armstrong, Neil A.

Science has not yet mastered prophecy. We predict too much for the next year and yet far too little for the next ten.

Address to Joint Sessions of Congress
September 16, 1969

Comte, Auguste

The aim of every science is foresight [*prévoyance*]. For the laws of established observation of phenomena are generally employed to foresee their succession. All men, however little advanced, make true predictions, which are always based on the same principle, the knowledge of the future from the past.

In Bertrand de Jouvenel
The Art of Conjecture (p. 111)

du Noüy, Pierre Lecomte

The aim of science is to foresee, and not, as has often been said, to understand. Science describes facts, objects and phenomena minutely, and tries to join them by what we call laws, so as to be able to predict events in the future.

Human Destiny
Chapter 2 (p. 13)

The aim of science is not so much to search for truth, or even truths, as to classify our knowledge and to establish relations between observable phenomena in order to be able to predict the future in a certain measure and to explain the sequence of phenomena in relation to ourselves.

Between Knowing and Believing
The Road to Reason (p. 188)

Hacking, Ian

Cutting up fowl to predict the future is, if done honestly and with as little interpretation as possible, a kind of randomization. But chicken guts are hard to read and invite flights of fancy or corruption.

The Emergence of Probability
Chapter 1 (p. 3)

Kaplan, Abraham

. . . if we can predict successfully on the basis of a certain explanation we have good reason, and perhaps the best sort of reason, for accepting the explanation.

The Conduct of Inquiry
Chapter IX, Section 40 (p. 350)

Russell, Bertrand

Science is the attempt to discover, by means of observation, and reasoning based upon it, first, particular facts about the world, and then laws connecting facts with one another and (in fortunate cases) making it possible to predict future occurrences.

Religion and Science
Grounds of Conflict (p. 8)

Unknown

To predict is one thing. To predict correctly is another.

Source unknown

PROBABLE ERROR

Russell, Bertrand

Who ever heard a theologian preface his creed, or a politician conclude his speech with an estimate of the probable error of his opinion?

In Edwin Hubble
The Nature of Science
The Nature of Science (p. 10)

PROBLEMS

Collingwood, R.G.
You can only solve a problem which you recognize to be a problem.

The New Leviathan
Part 1, Chapter 2, aphorism 2.66

Compton, Karl Taylor
Neither curiosity nor ingenuity is a modern impulse . . . The distinctive feature of science and technology at the present time is the accelerated pace of their development. This is partly due to continually improved techniques and organization, and it is partly due to the great accumulation of knowledge and art, because the more information and tools we have at our disposal, the more powerful can be the attack on any new problem.

A Scientist Speaks (pp. 1–2)

Condorcet, Jean
If a scholar poses himself a new problem, he can attack it fortified by the pooled resources of all his predecessors.

Eulogy for J. de Vaucanson before the Academy of Sciences
In Maurice Dumas
Scientific Instruments of the 17th and 18th Centuries and Their Makers (p. 119)

Einstein, Albert
Infeld, Leopold
The importance of a problem should not be judged by the number of pages devoted to it.

The Evolution of Physics
Preface (ix)

Flexner, Abraham
. . . science, in the very act of solving problems, creates more of them.

Universities
Chapter I, Section v (p. 19)

Frazier, A.W.

Often problems not solved earlier have not been posed earlier.

Hydrocarbon Processing
The Practical Side of Creativity
Volume 45, Number 1, January 1966

Fredrickson, A.G.

To be aware that a problem exists is the prerequisite for any attempt to solve the problem.

Chemical Engineering Education
The Dilemma of Innovating Societies (p. 148)
Summer 1969

Hilbert, David

As long as a branch of science offers an abundance of problems, so long it is alive; a lack of problems foreshadows extinction or the cessation of independent development.

Bulletin of the American Mathematical Society
Hilbert: Mathematical Problems (p. 438)
Volume 8, July 1902

Huxley, Julian

The time has gone by when the intelligent public needs to be reminded of the practical utility of science, or of the fact that the investigation of any problem, however apparently remote from everyday life, may be fraught with the most valuable consequences.

The Century Illustrated Monthly Magazine
Searching for the Elixir of Life (p. 629)
Volume 103, Number 4, February 1922

Pallister, William

Science solves life's problems, but she must solve them one-at-a-time. Her course and methods are evolutionary. She cannot solve insoluble problems; they must first become soluble. She grows, like every other plant, only from powdered fock, uses only the chemical constituents which are soluble.

Poems of Science
The Nature of Things (p. 14)

Rapoport, Anatol

. . . the problems scientists are called on to solve are for the most part selected by the scientists themselves. For example, our Department of Defense did not one day decide that it wanted an atomic bomb and then order the scientists to make one. On the contrary, it was Albert Einstein, a scientist, who told Franklin D. Roosevelt, a decision maker, that such a bomb was possible.

Science, Conflict and Society: Readings from Scientific American
The Use and Misuse of Games Theory (p. 286)

Shaw, George Bernard
. . . all problems are finally scientific problems.

The Doctor's Dilemma
Preface
The Technical Problem (p. lxxxiii)

Szent-Györgyi, Albert
Somehow, problems get into my blood and they don't give me peace, they torture me. I have to get them out of my system, and there is but one way to get them out—by solving them. A problem solved is no problem at all, it just disappears.

Perspectives in Biology and Medicine
On Scientific Creativity (p. 176)
Volume V, Number 2, Winter 1962

Thurstone, Louis Leon
Every scientific problem is a search for the relationship between variables. Every scientific problem can be stated most clearly if it is thought of as a search for the nature of the relation between two definitely stated variables. Very often a scientific problem is felt and stated in other terms, but it cannot be so clearly stated in any way when it is thought of as a function by which one variable is shown to be dependent upon or related to some other variable.

The Fundamentals of Statistics (p. 187)

Unknown
The only practical problem is what to do next.

Source unknown

If a problem has less than three variables it is not a problem. If it has more than eight, you cannot solve it.

Source unknown

Weil, Simone
Our science is like a store filled with the most subtle intellectual devices for solving the most complex problems, and yet we are almost incapable of applying the elementary principles of rational thought.

The Simone Weil Reader
The Power of Words (p. 271)

Wiesner, Jerome Bert
Some problems are just too complicated for rational logical solutions. They admit of insights, not answers.

In D. Lang
New Yorker
Profiles: A Scientist's Advice II
January 26, 1963

PROGRESS OF SCIENCE

Ardrey, Robert

The contemporary revolution in the natural sciences has proceeded in something more striking than silence. It has proceeded in secret. Like our tiny, furry, squirrel-like, earliest primate ancestors, seventy million years ago, the revolution has found obscurity its best defence and modesty the key to its survival. For it has challenged larger orthodoxies than just those of science, and its enemies exist beyond counting. From seashore and jungle, from ant-heap and travertine cave have been collected the inflammable materials that must some day explode our most precious myths. The struggle toward truth has proceeded, but as an underground intellectual movement seeking light under darkest cover.

African Genesis
Chapter I, Section 2 (p. 13)

Bernard, Claude

The progress of the experimental method consists in this,—that the sum of truths grows larger in proportion as the sum of errors grows less. But each one of these particular truths is added to the rest to establish more general truths. In this fusion, the names of promoters of science disappear little by little, and the further science advances, the more it takes an impersonal form and detaches itself from the past.

An Introduction to the Study of Experimental Medicine
Part I, Chapter II, Section iv (p. 42)

Bronowski, Jacob

The progress of science is the discovery at each step of a new order which gives unity to what had long seemed unlike.

Science and Human Values
The Creative Mind (p. 26)

Cohen, Morris R.

. . . the progress of science always depends upon our questioning the plausible, the respectably accepted, and the seemingly self-evident.

Reason and Nature
Book III
Chapter One, Section II (p. 348)

Coles, Abraham

Believing needless ignorance a crime,
You strive to reach the summit of your time;
To old age learning up from early youth
Your life one long apprenticeship to truth.
Wisely suspicious sometimes of the new,
Ye give alert acceptance to the true:
Even though it make old science obsolete,
It with a thousand welcomes still you greet . . .
Each Year adds something—many things ye know
Your sires knew not a Hundred Years ago.

The Microcosm and Other Poems
The Microcosm
Physician's Character and Aims—Science Progressive

Daly, Reginald Aldworth

Inasmuch as cosmogony and geology are both young sciences, consensus of opinions about the earth's origin and history is still reserved for the future. Meantime these sciences are advancing through the erection and testing of competing hypotheses; in other words, through speculation, controlled by all the available facts. Science progresses through systematic guessing in the good sense of the world.

Our Mobile Earth
Introduction (p. xx)

du Noüy, Pierre Lecomte

The scientist with imagination is the pioneer of progress.

The Road to Reason
Chapter 3 (p. 81)

Duclaux, Émile

It is because science is sure of nothing that it is always advancing.

In William Osler
Evolution of Modern Medicine
Chapter VI (p. 219)

Duhem, Pierre

Scientific progress has often been compared to a mounting tide; applied to the evolution of physical theories, this comparison seems to us very appropriate, and it may be pursued in further detail.

Whoever casts a brief glance at the waves striking a beach does not see the tide mount; he sees a wave rise, run, uncurl itself, and cover a narrow strip of sand, then withdraw by leaving dry the terrain which it had seemed to conquer; a new wave follows, sometimes going a little farther than the preceding one, but also sometimes not even reaching the sea shell made wet by the former wave. But under this superficial to-and-fro motion, another movement is produced, deeper, slower, imperceptible to the casual observer; it is a progressive movement continuing and coming of the waves is the faithful image of those attempts at explanation which arise only to be crumbled, which advance only to retreat; underneath there continues the slow and constant progress whose flow steadily conquers new lands, and guarantees to physical doctrines the continuity of a tradition.

The Aim and Structure of Physical Theory
Chapter III (pp. 38–9)

Einstein, Albert
I think that only daring speculation can lead us further and not accumulation of facts.

In Michele Besso
Correspondance 1903–1955
Letter to M. Besso
October 8, 1952 (p. 487)

Einstein, Albert
Infeld, Leopold
Science forces us to create new ideas, new theories. Their aim is to break down the wall of contradictions which frequently blocks the way of scientific progress. All the essential ideas in science were born in a dramatic conflict between reality and our attempts at understanding.

The Evolution of Physics
Quanta (p. 280)

France, Anatole
The progress of science renders useless the very books which have been the greatest aid to that progress. As those works are no longer useful, modern youth is naturally inclined to believe they never had any value; it despises them, and ridicules them if they happen to contain any superannuated opinion whatever.

The Crime of Sylvestre Bonnard
June 4 (p. 168)

Free, E.E.
Like a man on a bicycle science cannot stop; it must progress or collapse.

The World's Work
The Electrical Brains in the Telephone (p. 429)
Volume LIII, Number 4, February 1927

Heisenberg, Werner

Science progresses not only because it helps to explain newly discovered facts, but also because it teaches us over and over again what the word 'understanding' may mean.

Physics and Beyond
Chapter 10 (p. 124)

Huxley, Thomas

The rapid increase of natural knowledge, which is the chief characteristic of our age, is effected in various ways. The main army of science moves to the conquest of new worlds slowly and surely, nor ever cedes an inch of the territory gained. But the advance is covered and facilitated by the ceaseless activity of clouds of light troops provided with a weapon—always efficient, if not always an arm of precision—the scientific imagination. It is the business of these *enfants perdus* of science to make raids into the realm of ignorance wherever they see, or think they see, a chance; and cheerfully to accept defeat, or it may be annihilation, as the reward of error. Unfortunately the public, which watches the progress of the campaign, too often mistakes a dashing incursion of the Uhlans for a forward movement of the main body; fondly imagining that the strategic movement to the rear, which occasionally follows, indicates a battle lost by science. And it must be confessed that the error is too often justified by the effects of the irrepressible tendency which men of science share with all other sorts of men known to me, to be impatient of that most wholesome state of mind—suspended judgment; to assume the objective truth of speculations which, from the nature of the evidence in their favour, can have no claim to be more than working hypotheses.

Collected Essays
Volume VII
Man's Place in Nature
The Aryan Question and Prehistoric Man (p. 271–2)

Jacob, François

Contrary to what I once thought, scientific progress did not consist simply in observing, in accurately formulating experimental facts and drawing up a theory from them. It began with the invention of a possible world, or a fragment thereof, which was then compared by experimentation with the real world. And it was this constant dialogue between imagination and experiment that allowed one to form an increasingly fine-grained conception of what is called reality.

In William Calvin
The Cerebral Symphony (p. 206)

Kuhn, Thomas S.

. . . we must explain why science—our surest example of sound knowledge—progresses as it does, and we first must find out how, in fact, it does progress.

In Imre Lakatos and Alan Musgrave
Criticism and the Growth of Knowledge
Falsification and the Methodology of Scientific Research Programmes (p. 20)

Mayr, Ernst

Progress in science is achieved in two ways: through new discoveries, such as x-rays, the structure of DNA, and gene splicing, and through the development of new concepts, such as the theories of relativity, of the expanding universe, of plate tectonics, and of common descent. Among all the new scientific concepts, perhaps none has been as revolutionary in its impact on our thinking as Darwin's theory of natural selection.

Toward A New Philosophy of Biology (p. 95)

Medawar, Sir Peter

To deride the hope of progress is the ultimate fatuity, the last word in poverty of spirit and meanness of mind.

The Hope of Progress
Introduction (p. 1)

It can be said that Science today progresses only by peeling away, one after another, all the coverings of apparent stability in the world; disclosing beneath the immobility of the infinitely small, movement of extra rapidity, and beneath the immobility of the Immense, movement of extra slowness.

The Future of Man
Some Reflections on Progress (p. 62)

Planck, Max

It is a rather zigzag pattern that the curve of scientific progress follows; indeed I might say that the forward movement is of an explosive type, where the rebound is an attendant characteristic of the advance. Every applied hypothesis which succeeds in throwing the searchlight of a new vision across the field of physical science represents a plunge into the darkness; because we cannot at first reduce the vision to a logical statement. Then follows the birth-struggle of a new theory. Once this has seen the light of day it has to go forward willy-nilly until the stamp of its destiny is put on it when the test of the research measurements is applied.

Where Is Science Going?
Nature's Image in Science (pp. 90–1)

Popper, Karl R.

In science it would be a tremendous loss if we were to say: "We are not making very much progress. Let us sweep away all science and start afresh." The rational procedure is to correct it and to revolutionize it, but not to sweep it away. You may create a new theory, but the new theory is created in order to solve those problems which the old theory did not solve.

Conjectures and Refutations
Chapter 4 (p. 132)

Priestley, Joseph

If the progress continues the same in another period, of equal length, what a glorious science shall we see unfold, what a fund of entertainment is there in store for us, and what important benefits must derive mankind.

Quoted by John G. McEvoy
The British Journal for the History of Science
Electricity, Knowledge, and the Nature of Progress
in Priestley's Thought (p. 76)
Volume XII, Number 40, 1979

Richet, Charles

One can only progress in the sciences—with the exception of Mathematics—at the price of great pecuniary sacrifice.

The Natural History of a Savant
Chapter II (p. 21)

All progress in science is progress in civilization, and consequently contributes to the welfare of man.

The Natural History of a Savant
Chapter XIII (p. 145)

Romanoff, Alexis

Science is the key to the progress of the world.

Encyclopedia of Thoughts
Aphorism 20

Thomson, Sir George

. . . the progress of science is a little like making a jig-saw puzzle. One makes collections of pieces which certainly fit together, though at first it is not clear where each group should come in the picture as a whole, and if at first one makes a mistake in placing it, this can be corrected later without dismantling the whole group.

The Inspiration of Science
Introduction (pp. 5–6)

Whitehead, Alfred North

The progress of science consists in observing interconnections and in showing with a patient ingenuity that the events of this ever-shifting world are but examples of a few general connexions or relations, called laws. To see what is general in what is particular, and what is permanent in what is transitory, is the aim of scientific thought.

An Introduction to Mathematics
Chapter 1 (p. 4)

. . . It is true that you have missed everything of importance, but you have hit upon the method . . .
Sir Arthur Conan Doyle – (See p. 132)

PROOF

Bell, Eric T.

There is a sharp disagreement among competent men as to what can be proved and what cannot be proved, as well as an irreconcilable divergence of opinions as to what is sense and what is nonsense.

Debunking Science (p. 18)

Bernal, J.D.

. . . the cardinal rule in science is that a statement must be provable—but that does not mean that it has to be proved *now*.

In S.W. Fox (Editor)
The Origins of Prebiological Systems and Their Molecular Matrices
The Folly of Probability
Discussion (pp. 53–5)

Blake, William

What is now proved was once only imagined.

The Complete Writings of William Blake
The Marriage of Heaven and Hell
L. 13

Davis, Philip J.
Hersh, Reuben

Proof serves many purposes simultaneously . . . Proof is respectability. Proof is the seal of authority.

Proof, in its best instance, increases understanding by revealing the heart of the matter.

Proof suggests new mathematics . . . Proof is mathematical power, the electric voltage of the subject which vitalizes the static assertions of the theorems.

The Mathematical Experience (p. 151)

Eddington, Sir Arthur Stanley

Proof is an idol before whom the pure mathematician tortures himself.

The Nature of the Physical World
Chapter XV (p. 337)

Hilbert, David

. . . it is an error to believe that rigor in the proof is the enemy of simplicity.

Bulletin of the American Mathematical Society
Hilbert: Mathematical Problems (p. 441)
Volume 8, 2nd series, October 1901–July 1902

Hoyle, Fred

What constitutes proof in one generation is not the same thing as proof in another.

Of Men and Galaxies (pp. 16–7)

Manin, Yu.I.

. . . a good proof is one which makes us wiser.

A Course in Mathematical Logic
Chapter II, Section 4 (p. 51)

Proverb, English

The proof of the pudding is in the eating.

Source unknown

Truzzi, Marcello

And when such claims are extraordinary, that is, revolutionary in their implications for established scientific generalizations already accumulated and verified, we must demand extraordinary proof.

Zetetic Scholar
Editorial
Volume 1, Number 1, Fall/Winter 1976 (p. 4)

Unknown

Sex and drugs? They're nothing compared with a good proof!

Mathematical and Scientific Quotes from Cambridge
The Internet

An obscure proof which I managed to present in an obscure way.

Source unknown

Zemanian, Armen H.

The usual techniques for proving things are often inadequate because they are merely concerned with truth. For more practical objectives, there are other powerful—but generally unacknowledged—methods.

Proof of Blatant Assertion:
Use words and phrases like "clearly . . .", "obviously . . .", "it is easily shown that . . .", and "as any fool can plainly see . . ."

Proof by Seduction:
"If you will just agree to believe this, you might get a better final grade."

Proof by Intimidation:
"You better believe this if you want to pass the course."

Proof by Interruption:
Keep interrupting until your opponent gives up.

Proof by Misconception:
An example of this is the Freshman's Conception of the Limit Process: "2 equals 3 for large values of 2." Once introduced, any conclusion is reachable.

Proof by Obfuscation:
A long list of lemmas is helpful in this case—the more, the better.

Proof by Confusion:
This is a more refined form of proof by obfuscation. The long list of lemmas should be arranged into circular patterns of reasoning—and perhaps more baroque structures such as figure-eights and fleurs-de-lis.

Proof by Exhaustion:
This is a modification of an inductive proof. Instead of going to the general case after proving the first one, prove the second case, then the third, then the fourth, and so on—until a sufficiently large n is achieved whereby the nth case is being propounded to a soundly sleeping audience.

The Physics Teacher
Appropriate Proof Techniques (p. 287)
Volume 32, Number 5, May 1994

PROPOSITIONS

Russell, Bertrand

The belief or unconscious conviction that all propositions are of the subject–predicate form—in other words, that every fact consists in some thing having some quality—has rendered most philosophers incapable of giving any account of the world of science and daily life.

Our Knowledge of the External World
Lecture II (pp. 54–5)

PUBLICATIONS

Batchelor, G.K.

Reading a paper is a voluntary and demanding task, and a reader needs to be enticed and helped and stimulated by the author.

Journal of Fluid Mechanics
Preoccupations of a Journal Editor
Volume 106, 1981

Writing up work for publication is an occasional and unfamiliar task needing skills which are not otherwise cultivated. As if this were not handicap enough, one sometimes also hears misguided advice—such as: aim at conciseness above all else, avoid explanatory interpolations, never use the first person—from people who suppose that a standardized telex style is essential for clarity.

Journal of Fluid Mechanics
Preoccupations of a Journal Editor
Volume 106, 1981

Bernal, J.D.

The very bulk of scientific publications is itself delusive. It is of very unequal value; a large proportion of it, possibly as much as three-quarters, does not deserve to be published at all, and is only published for economic considerations which have nothing to do with the real interests of science . . . Publication is often premature and dictated by the need of establishing priorities.

The kind of organisation we wish to aim at is one in which all relevant information should be available to each research worker and in amplitude proportional to its degree of relevance. Further, that not only should the information be available, but also that it should be to a large extent put at the disposal of the research worker without his having to take any steps to get hold of it.

The Social Function of Science
Chapter V
Scientific Publications (p. 118)

de Saint-Exupéry, Antoine

. . . and if the recollections of any one among them seem interesting to him, the geographer orders an inquiry into that explorer's moral character.

The Little Prince
XV (p. 53)

Galton, Francis

When apples are ripe, a trifling event suffices to decide which of them shall first drop off its stick; so a small accident will often determine the scientific man who shall first make and publish a new discovery.

Nature
Scientific Achievement and Aptitude (p. 686)
Volume 118, Number 2976, November 13, 1926

Lofting, Hugh

What to leave out and what to put in? That's the problem.

Doctor Dolittle's Zoo

Watson, David Lindsay

Professional scientists tacitly assume that the chief operations by which science is created are those which are performed before the footlights, in the laboratory or the study, and recorded so impressively in scientific publications.

Scientists Are Human
Introduction (p. xiii)

Wilson, E. Bright, Jr.

A large number of incorrect conclusions are drawn because the possibility of chance occurrences is not fully considered. This usually arises through lack of proper controls and insufficient repetitions. There is the story of the research worker in nutrition who had published a rather surprising conclusion concerning rats. A visitor asked him if he could see more of the evidence. The researcher replied, 'Sure, there's the rat.'

An Introduction to Scientific Research
Chapter 3, Section 10 (p. 34)

QUESTIONS

Bloor, David
To ask questions of the sort which philosophers address to themselves is usually to paralyse the mind . . .

Knowledge and Social Imagery
Chapter 5 (p. 45)

Bohm, David
. . . it is frequently realised that half the battle is over when we know what are the right questions to ask.

British Journal for the Philosophy of Science
On the Relationship between Methodology in Scientific Research and the Content of Scientific Knowledge (p. 105)
Volume XII, Number 46, August 12, 1961

Boltzmann, Ludwig
If a general intends to conquer a hostile city, he will not consult his map for the shortest road leading there; rather he will be found to make the most various detours, and every hamlet, even if quite off the path, will become a valuable point of leverage for him, if only he can take it; impregnable places will be isolated. Likewise, the scientist asks not what are the currently most important questions, but, "Which are at present solvable?", or sometimes simply, "In which can we make some small but genuine advance?" As long as the alchemist merely sought the philosopher's stone and aimed at finding the art of making gold, all their endeavors were fruitless; it was only when people restricted themselves to seemingly less valuable questions that they created chemistry. Thus natural science appears completely to lose from sight the large and general questions; but all the more splendid is the success when, groping in the thicket of special questions, we suddenly find a small opening that allows a hitherto undreamt of outlook on the whole.

Theoretical Physics and Philosophical Problems
The Second Law of Thermodynamics (p. 13–4)

Bromberger, Sylvain

A clear mark of scientific genius is the ability to see certain well-known facts as departures from general rules . . . and the germane ability to ask why-questions that occur to no one else.

In Robert G. Colodny
Mind and Cosmos
Why-Questions (p. 103)

Colby, Frank Moore

Every man ought to be inquisitive through every hour of his great adventure down to the day when he shall no longer cast a shadow in the sun. For if he dies without a question in his heart, what excuse is there for his continuance?

In Hans Selye
From Dream to Discovery
Why Should You Do Research (p. 10)

Fischer, D.H.

Questions are the engines of intellect, the cerebral machines which convert energy to motion, and curiosity to controlled inquiry.

Historians' Fallacies
Chapter I (p. 3)

Hoffer, Eric

To spell out the obvious is often to call it in question.

The Passionate State of Mind
Number 220

Little, T.M.

The purpose of an experiment is to answer questions. The truth of this seems so obvious, that it would not be worth emphasizing were it not for the fact that the results of many experiments are interpreted and presented with little or no reference to the questions that were asked in the first place.

Hortscience
Interpretation and Presentation of Results (pp. 637–40)
Volume 16, 1981

Midgley, Mary

The astonishing successes of western science have not been gained by answering every kind of question, but precisely by refusing to. Science has deliberately set narrow limits to the kinds of questions that belong to it, and further limits to the questions peculiar to each branch. It has practiced an austere modesty, a rejection of claims to universal authority.

New Scientist
Can Science Save Its Soul? (p. 25)
August 1, 1992

Popper, Karl R.

Never let yourself be goaded into taking seriously problems about words and their meanings. What must be taken seriously are questions of fact, and assertions about facts: theories and hypotheses; the problems they solve; and the problems they cause.

Unended Quest
7 (p. 19)

Russell, Bertrand

Clearly our first problem must be to define the issue, since nothing is more prolific of fruitless controversy than an ambiguous question.

Proceedings of the University of Durham Philosophical Society
Determinism and Physics, 1936

Seignobos, Charles

It is useful to ask oneself questions, *but very dangerous to answer them*.

In Marc Bloch
The Historian's Craft
Introduction (p. 17)

Wittgenstein, Ludwig

As long as I continue to come across questions in more remote regions which I can't answer, it is understandable that I should still not be able to find my way around regions that are less remote. For how do I know that what stands in the way of an answer here is not precisely what is preventing me from clearing away the fog over there?

Culture and Value (p. 66e)

REALITY

Born, Max
The simple and unscientific man's belief in reality is fundamentally the same as that of the scientist.

Physics in My Generation
On the Meaning of Physical Theories (p. 16)

Bronowski, Jacob
Reality is not an exhibit for man's inspection, labeled "Do not touch." There are no appearances to be photographed, no experiences to be copied, in which we do not take part. Science, like art, is not a copy of nature but a re-creation of her.

Science and Human Values
The Creative Mind (p. 20)

Burtt, E.A.
Man begins to appear for the first time in the history of thought as an irrelevant spectator and insignificant effect of the great mathematical system which is the substance of reality.

The Metaphysical Foundations of Modern Physical Science (p. 80)

Olson, Sigurd F.
Flashes of insight or reality are sunbursts of the mind.

Reflections From the North Country
Flashes of Insight (p. 131)

Pagels, Heinz
We may begin to see reality differently simply because the computer . . . provides a different angle on reality.

The Dreams of Reason: The Computer and the Rise of the Sciences of Complexity
Preface (p. 13)

Smith, David

Everything imagined is reality. The mind cannot conceive unreal things.

Vogue
The Private Thoughts of David Smith (p. 198)
November 15, 1968

Trilling, Lionel

In the American metaphysic, reality is always material reality, hard, resistant, unformed, impenetrable, and unpleasant.

The Liberal Imagination
Reality in America (pp. 10–1)

Unknown

The difference between reality and theory is that reality takes into account all of the theory.

Source unknown

Walgate, Robert

. . . what the scientist must now admit is that in many problems of great consequence to people reality may not be accessible, in practice, through entirely manipulative and analytical methods.

New Scientist
Breaking Through the Disenchantment (p. 667)
September 18, 1975

Whitehead, Alfred North

Progress in truth—truth of science and truth of religion—is mainly a progress in the framing of concepts, in discarding artificial abstractions or partial metaphors, and in evolving notions which strike more deeply into the root of reality.

Religion in the Making
Truth and Criticism (p. 117)

REASON

Beck, Lewis White

In the logic of science there is a principle as important as that of parsimony: it is that of sufficient reason.

The Scientific Monthly
The "Natural Science Ideal" in the Social Sciences (p. 393)
Volume LXVIII, June 1949

Brophy, B.

Reason is necessarily the language of moral, political, and scientific argument: not because reason is holy or on some elevated plane, but because it *isn't*; because it is accessible to all humans; because, as well as working, it can be seen to work.

In Stanley and Rosiland Godlovitch and John Harris (Editors)
Animals, Men and Morals
In Pursuit of Fantasy (p. 126)

Browne, Sir Thomas

Every man's own reason is his best Oedipus.

Religio Medici
Part I, Section 6

Burton, Sir Richard

Reason is Life's sole arbiter, the magic Laby'rinth's single clue . . .

The Kasidah of Hâjî Abdû El-Yezdî
Part vii, stanza xxii

Congreve, William

. . . error lives
Ere reason can be born.

The Mourning Bride
Act III, Scene 1

Drummond, Sir William

. . . he, who will not reason, is a bigot; he, who cannot, is a fool; and he, who dares not, is a slave.

Academical Questions
Volume I
Preface (p. xv)

Galileo Galilei

In science the authority embodied in the opinion of thousands is not worth a spark of reason in one man.

In Pedro Redondi
Galileo: Heretic
Chapter 2 (p. 37)

Gay-Lussac, Joseph Louis

The scientific glory of a country may be considered in some measure, as an indication of its innate strength. The exaltation of Reason must necessarily be connected with the exaltation of the other faculties of the mind; and there is one spirit of enterprize, vigour and conquest in science, arts, and arms.

In Maurice Crosland
Gay-Lussac: Scientist and Bourgeois
Chapter 4 (p. 80)

Kant, Immanuel

Mathematics and *physics* are the two theoretical sciences of reason, which have to determine their objects *a priori*.

The Critique of Pure Reason
Preface to the Second Edition (p. 5)

Osler, Sir William

With reason, science never parts company, but with feeling, emotion, passion, what has she to do? They are not of her; they owe her no allegiance. She may study, analyze, and define, she can never control them, and by no possibility can their ways be justified to her.

Aequanimitas
The Leaven of Science (p. 93)

Shakespeare, William

His reasons are as two grains of wheat hid in two bushels of chaff: you shall seek all day ere you find them, and when you have them, they are not worth the search.

The Merchant of Venice
Act I, Scene 1, L. 115

Good reason must, of force, give place to better.

Julius Caesar
Act IV, Scene 3, L. s03

RELATIONS

Buchanan, Scott
Science is an allegory that asserts that the relations between the parts of reality are similar to the relations between terms of discourse.

Poetry and Mathematics
Chapter IV (p. 104)

Dingle, Herbert
. . . if, as we must surely do, we wish to characterize science by the elements in it that persist and grow, and not by that which continually changes, we must recognize . . . the progressive discovery of relations between the various constituents of our experience, . . . Amid all the changes of theories and pictures and conceptions, the relations remain and steadily accumulate. Franklin found that lightning was a manifestation of the electric ether revealed in laboratory experiments. The electric ether has disappeared, and other theories of electricity have in turn succeeded it and disappeared also, but the relation between lightning and laboratory sparks remains. Maxwell established a relation between light and electromagnetic oscillations. His ether also has gone, but the relation stays. All permanent advances in science are discoveries of relations between phenomena, and the factor in science that shows a steady uninterrupted growth is the extent of the field of related observations.

The Scientific Adventure
Chapter One (p. 40)

Durant, William
Science tells us how to heal and how to kill; it reduces the death rate in retail and then kills us wholesale in war; but only wisdom—desire coordinated in the light of all experience—can tell us when to heal and when to kill. For a fact is nothing except in relation to a purpose and a whole.

The Story of Philosophy
Introduction (p. 3)

Keyser, Cassius J.
To be is to be related.

Mole Philosophy & Other Essays
Mathematics and Man (p. 94)

RESEARCH

Bates, Marston

Research is the process of going up alleys to see if they are blind.

In Jefferson Hane Weaver
The World of Physics
Volume 2 (p. 63)

Belloc, Hilaire

. . . anyone of common mental and physical health can practise scientific research . . . Anyone can try by patient experiment what happens if this or that substance be mixed in this or that proportion with some other under this or that condition. Anyone can vary the experiment in any number of ways. He that hits in this fashion on something novel and of use will have fame . . . The fame will be the product of luck, and industry. It will not be the product of special talent.

Essays of a Catholic
Science as the Enemy of Truth (pp. 226–7)

Bradley, A.C.

Research, though toilsome, is easy; imaginative vision, though delightful, is difficult.

Oxford Lectures on Poetry
Shakespeare's Theatre and Audience (p. 362)

Brown, J. Howard

A man may do research for the fun of doing it but he cannot expect to be supported for the fun of doing it.

Journal of Bacteriology
The Biological Approach to Bacteriology (p. 9)
Volume XXIII, Number 1, January 1932

Browning, Robert

. . . As is your sort of mind,
So is your sort of search: you'll find
What you desire.

The Poems
Easter Day
Part vii, L. 3 (p. 501)

Bunge, Mario

Most scientists are prepared to grant that the chief theoretical (that is, nonpragmatic) aim of scientific research is to answer, in an intelligible, exact, and testable way, five kinds of questions, namely those beginning with *what* (or *how*), *where*, *when*, *whence*, and *why* . . . [T]he Five W's of Science. (Only radical empiricists deny that science has an explanatory function, and restrict the task of scientific research to the description and prediction of observable phenomena.) Also, most scientists would agree that all five W's are gradually (and painfully) being answered through the establishment of scientific laws, that is, general hypotheses about the patterns of being and becoming.

Causality: The Place of the Causal Principle in Modern Science
Chapter 10 (p. 248)

Bush, Vannevar

Basic research leads to new knowledge. It provides scientific capital. It creates the fund from which the practical applications of knowledge must be drawn. New products and new processes do not appear full-grown. They are founded on new principles and new conceptions, which in turn are painstakingly developed by research in the purest realms of science.

Endless Horizons (pp. 52–3)

Chesterton, G.K.

Research is the search of people who don't know what they want.

The G.K. Chesterton Calendar
May Twenty-fifth

da Vinci, Leonardo

Nothing is written as the result of new researches.

Leonardo da Vinci's Notebooks (p. 53)

George, William H.

Scientific research is not itself a science: it is still an art or craft.

The Scientist in Action
Four Qualities of Scientific Research (p. 29)

Green, Celia

The way to do research is to attack the facts at the point of greatest astonishment.

The Decline and Fall of Science
Aphorisms (p. 1)

Research is a way of taking calculated risks to bring about incalculable consequences.

The Decline and Fall of Science
Aphorisms (p. 1)

Gregg, Alan

One wonders whether the rare ability to be completely attentive to, and to profit by, nature's slightest deviation from the conduct expected of her is not the secret of the best research minds and one that explains why some men turn to most remarkably good advantage seemingly trivial accidents. Behind such attention lies an unremitting sensitivity . . .

The Furtherance of Medical Research
The Medical Research Worker (p. 98)

Harris, Ralph

RESEARCH: A procedure for expressing your political prejudices in convincing statistical guise.

Growth, Advertising, and the Consumer
Everyman's Guide to Contemporary Economic Jargon (p. 24)

Jevons, William Stanley

So-called original research is now regarded as a profession, adopted by hundreds of men, and communicated by a system of training.

The Principles of Science
Chapter XXVI (p. 574)

Johnson, Harry G.

To an important extent, indeed, scientific research has become the secular religion of materialistic society; and it is somewhat paradoxical that a country whose constitution enforces the strict separation of church and state should have contributed so much public money to the establishment and propagation of scientific pessimism.

In National Academy of Sciences
Basic Research and National Goals:
A Report to the Committee on Science and Astronautics
Federal Support of Basic Research: Some Economic Issues
Note 4 (p. 141)

Kettering, Charles F.
We find that in research a certain amount of intelligent ignorance is essential to progress; for if you know too much, you won't try the thing.

In T.A. Boyd
Professional Amateur, The Biography of Charles Franklin Kettering (p. 106)

Kettering, Charles F.
Smith, Beverly
[Research] may use a laboratory or it may not. It is purely a principle, and everybody can apply it in his own life. It is simply a way of trying to find new knowledge and ways of improving things which you are not satisfied with.

The American Magazine
Ten Paths to Fame and Fortune (p. 14)
December 1937

Matthew 7:7
Seek and ye shall find; knock, and it shall be opened unto you.

The Bible

Medawar, Sir Peter
If politics is the art of the possible, research is surely the art of the soluble. Both are immensely practical-minded affairs.

New Statesman
Summer Books (p. 950)
June 19, 1964

Richet, Charles
The gift for investigation appears at an early age: *the demon of research* speaks to men whilst they are still young.

The Natural History of a Savant
Chapter VI (pp. 38–9)

Understand this clearly; that the right method, even for obtaining a useful practical result, is not to worry about the practice, but to concentrate intensely on pure investigation, without being hampered by any parasitic considerations other than whatever conduces to greater facility for research.

The Natural History of a Savant
Chapter XII (p. 134)

Romanoff, Alexis
Scientific research provides the shortest route to useful practice.

Encyclopedia of Thoughts
Aphorisms 112

Scientific research is based chiefly on creative thinking.

Encyclopedia of Thoughts
Aphorisms 219

Schild, A.
If one can tell ahead of time what one's research is going to be, the research problem cannot be very deep and may be said to be almost nonexistent.

Canadian Association of University Teachers
On the Matter of Freedom: The University and the Physical Sciences
Bulletin
Volume 11, Number 4, 1963

Smith, Homer W.
On every scientist's desk there is a drawer labeled UNKNOWN in which he files what are at the moment unsolved questions, lest through guesswork or impatient speculation he come upon incorrect answers that will do him more harm than good. Man's worst fault is opening the drawer too soon. His task is not to discover final answers but to win the best partial answers that he can, from which others may move confidently against the unknown, to win better ones.

From Fish to Philosopher
Chapter 13 (p. 231)

Smith, Theobald
. . . it is the care we bestow on apparently trifling, unattractive and very troublesome minutiae which determines the results.

New York Medical Journal
Volume lii, 1890 (p. 485)

The joy of research must be found in doing, since every other harvest is uncertain.

Journal of Bacteriology
Letter from Dr. Theobald Smith (p. 20)
Volume XXVII, Number 1, January 1934

Terence
Nothing is so difficult but that it may be found out by seeking.

Heauton Timorumenos
Act IV, Scene 2, L. 675

Thomson, J.J.
. . . research in applied science leads to reforms, research in pure science leads to revolutions . . .

The Life of Sir J.J. Thomson (p. 199)

Wells, H.G.

The whole difference of modern scientific research from that of the Middle Ages, the secret of its immense success, lies in its collective character, in the fact that every fruitful experiment is published, every new discovery of relationships explained.

New Worlds for Old
Chapter II (p. 22)

Whitney, W.R.

The valuable attributes of research men are conscious ignorance and active curiosity.

Science
The Stimulation of Research in Pure Science which has Resulted from the Needs of Engineers and of Industry (p. 289)
Volume LXV, Number 1862, March 25, 1927

Willstätter, Richard

It is the example of doing research which trains researchers, and that is how the academic career is organized. I admit that H.G. Wells says, in his novel *Marriage*, "The trained investigator is quite the absurdest figure in the farce of contemporary intellectual life; he is like a Bath chair perpetually starting to cross the Himalayas by virtue of a licence to do so. For such enterprises one must have wings. Organization and genius are antipathetic." This is more likely to be true in art. In research, the great achievements rarely come from unschooled youthful geniuses. There is little prospect that a beginner with an original mind or even one with the gift of genius will be able to scale the heights unless a mature leader sets him a daily example of steadiness and perseverance, devotion and unselfishness as self-evident characteristics of a scientist.

From My Life
Chapter 11 (p. 344)

Wordsworth, William

Lost in the gloom of uninspired research.

The Poetical Works of William Wordsworth
The Excursion
Despondency Corrected, L. 626

RULES

Arnheim, Rudolf
An orgy of self-expression is no more productive than blind obedience to rules.

Art and Visual Perception
Introduction (p. vii)

Edison, Thomas
There ain't no rules around here! We're trying to accomplish something!

In Robert Byrne
The Fourth, and By Far the Most Recent, 637 Best Things Anybody Ever Said

Galsworthy, John
KEITH: . . . I don't see the use in drawin' hard and fast rules. You only have to break 'em.

Plays
Second Series
Eldest Son
Act I, Scene II (p. 13)

Norton, Robert
. . . every Art hath certain Rules and Principles, . . . without the knowledge of which no man can attain unto a necessary perfection for practice thereof . . .

The Gunner
The Preface to the courteous Readers (second page)

Wilson, John
. . . the Exception proves the Rule.

In Milton C. Nahm (Editor)
The Cheats
To the Reader
L. 27 (p. 236)

SAVANT

Richet, Charles
For the savant, science must be a religion. Everything that is discovered, be it great or small, has its origin in this faith.

The Natural History of a Savant
Chapter VI (p. 47)

Without the sciences man would rank below the beasts.
de la Rivière Mercer – (See p. 228)

SCIENCE

Adams, Henry
No sand-blast of science had yet skimmed off the epidermis of history, thought, and feeling.

The Education of Henry Adams
Rome (p. 90)

Akenside, Mark
Speak, ye, the pure delight, whose favour'd steps
The lamp of science, through the jealous maze
Of nature guides, when haply you reveal
Her secret honours . . .

The Poems of the Pleasures
The Pleasures of Imagination
Part II (p. 75)

Appleyard, Bryan
Science is not a neutral or innocent commodity which can be employed as a convenience . . . Rather it is spiritually corrosive, burning away at ancient authorities and traditions. It has shown itself unable to coexist with anything.

Understanding the Present: Science and the Soul of Modern Man
Chapter 1 (p. 9)

Aristotle
Science is a mode of conceiving universal and necessary truths.

Nicomachean Ethics
vi, VI, 1

Ayala, Francisco J.
Science is systematic organisation of knowledge about the universe on the basis of explanatory hypotheses which are genuinely testable. Science advances by developing gradually more comprehensive theories; that is, by formulating theories of greater generality which can account for

observational statements and hypotheses which appear as *prima facie* unrelated.

Studies in the Philosophy of Biology
Introduction (p. ix)

Bacon, Francis

Morever the works already known are due to chance and experiment rather than to sciences; for the sciences we now possess are merely systems for the nice ordering and setting forth of things already invented; not methods of invention or directions for new works.

Norvum Organum
Aphorism VIII

Science, being the wonder of the ignorant and unskillful, may be not absurdly called a monster. In figure and aspect it is represented as many-shaped, in allusion to the immense variety of matter with which it deals. It is said to have the face and voice of a woman, in respect of its beauty and facility of utterance. Wings are added because the sciences and the discoveries of science appeared and fly-aboard in an instant; the communication of knowledge being like that of one candle with another, which lights up at once. Claws, sharp and hooked, are ascribed to it with great elegance, because the axioms and arguments of science penetrate and hold fast the mind, so that it has no means of evasion or escape.

Selected Writings of Francis Bacon
Sphinx on Science (pp. 418–9)

The divisions of the sciences are not like different lines that meet in one angle, but rather like the branches of trees that join in one trunk.

In J.A. Thomson
Introduction to Science
Chapter IV (p. 92)

Balfour, A.J.

Science preceded the theory of science, and is independent of it. Science preceded naturalism, and will survive it.

The Foundations of Belief
II

Beard, Charles A.

. . . science can discover the facts that condition realization and furnish instrumentalities for carrying plan and purpose into effect. Science without dreams is sterile. Dreams without research and science are empty. The deed of ignorance is perilous; deedless information is futile. Unlimited, idea and deed may create a civilization.

A revolution in thought is at hand, a revolution as significant as the Renaissance: the subjection of science to ethical and esthetic purpose.

Hence the next great survey undertaken in the name of the social sciences may begin boldly with a statement of values agreed upon, and then utilize science to discover the conditions, limitations, and methods involved in realization.

Social Forces
Limitations to the Application of Social Science Implied in Recent Social Trends (p. 510)
Volume XI, Number 4, May 1933

Beattie, James

'Twas thus by the glare of false science betray'd,
That leads, to bewilder; and dazzles, to blind; . . .

The Complete Poetical Works of Gray, Beattie, Blair, Collins, Thomson, and Kirke White
The Hermit
Stanza 5

Bernal, J.D.

Science is one of the most absorbing and satisfying pastimes, and as such it appeals in different ways to different types of personality. To some it is a game against the unknown where one wins and no one loses, to others, more humanly minded, it is a race between different investigators as to who should first wrest the prize from nature. It has all the qualities which make millions of people addicts of the crossword puzzle or the detective story, the only difference being that the problem has been set by nature or chance and not by man, that the answers cannot be got with certainty, and when they are found often raise far more questions than the original problem.

The Social Function of Science (p. 97)

Boas, George

Science is the art of understanding nature.

In Laurence M. Gould
UNESCO Courier
Science and the Culture of Our Times (p. 6)
February 1968

Bonaparte, Napoleon

The sciences, which have revealed so many secrets and destroyed so many prejudices, are destined to render us yet greater service. New truths, new discoveries will unveil secrets still more essential to the happiness of men—but only if we give our esteem to the scientists and our protection to the sciences.

In J. Christopher Herold
The Mind of Napoleon
Science and the Arts (p. 135)

Bronowski, Jacob
[Science] does not watch the world, it tackles it.

The Common Sense of Science
Chapter VII, Section 4 (p. 104)

The world today is made, it is powered by science; and for any man to abdicate an interest in science is to walk with open eyes towards slavery.

Science and Human Values
The Creative Mind (p. 13)

All science is the search for unity in hidden likenesses.

Science and Human Values
The Creative Mind (p. 13)

The discoveries of Science, the works of art are explorations—more, are explosions, of a hidden likeness.

Science and Human Values
The Creative Mind (p. 19)

The values of science derive neither from the virtues of its members, nor from the finger-wagging codes of conduct by which every profession reminds itself to be good. They have grown out of the practice of science, because they are the inescapable conditions for its practice.

Science and Human Values
The Sense of Human Dignity (p. 60)

Science is the creation of concepts and their exploration in the facts. It has no other test of the concept than its empirical truth to fact.

Science and Human Values
The Sense of Human Dignity (p. 60)

Science is not a mechanism but a human progress, and not a set of findings but the search for them.

Science and Human Values
The Sense of Human Dignity (p. 63)

Science has nothing to be ashamed of even in the ruins of Nagasaki. The shame is theirs who appeal to other values than the human imaginative values which science has evolved.

Science and Human Values
The Sense of Human Dignity (pp. 73)

That is the essence of science: ask an impertinent question, and you are on the way to a pertinent answer.

The Ascent of Man
Chapter 4

Science is a great many things, . . . but in the end they all return to this: science is the acceptance of what works and the rejection of what does not. That needs more courage than we might think.

It needs more courage than we have ever found when we have faced our worldly problems.

The Common Sense of Science (p. 148)

Brooks, H.

. . . the only definite answer that can be given to these questions lies in the nature of science as a system of acquiring and validating knowledge. Science—especially natural science—has a public characteristic that is still lacking in other forms of knowledge. The results of scientific research have to stand the scrutiny of a large and critical scientific community, and after a time those that stand the test tend to be accepted by all literate mankind. Outside the scientific community itself this acceptance tends to be validated by the practical results of science. If it works it must be true. There is no question that the successful achievement of an atom bomb provided a certain intellectual validation for nuclear physics, quite apart from its practical value. Part of the public character of science results from the fact that it is always in principle subject to independent validation or verification. It is like paper money that can always be exchanged for gold or silver on demand. Just because everybody believes that he can get gold for paper, nobody tries; so the public seldom questions the findings of science, just because it believes that they can always be questioned and validated on demand. This is much less true of other forms of knowledge and culture, which may be of equal social importance but are more subjective and more dependent on the vagaries of private tastes and value systems.

In National Academy of Sciences
Basic Research and National Goals:
A Report to the Committee on Science and Astronautics
Future Needs for the Support of Basic Research (pp. 85–6)

Bube, R.H.

If you can't see it, hear it, feel, taste, or smell it, then science can't work with it.

The Encounter between Science and Christianity
Chapter 1 (p. 18)

Buckham, John Wright

. . . the mind that has been trained simply or predominately in Science is an unconsciously meager and ill-furnished mind. The range of its interests is mainly technical and specialized. To look into a mind of this type is like looking into a laboratory. It is excellent as a workshop, but there are no pictures on the walls, no books, no flowers . . . What are the resources of such a mind, its points of contact with human-kind?

The Century Illustrated Monthly Magazine
The Passing of the Scientific Era (p. 435)
August 1929

Bulwer-Lytton, Edward

. . . science is not a club, it is an ocean; it is open to the cockboat as the frigate. One man carries across it a freightage of ingots, another may fish there for herrings. Who can exhaust the sea? who say to intellect, 'the deeps of philosophy are preoccupied?'

The Caxtons
Book IV, III (p. 96)

Bunge, Mario

The motto of science is not just *Pauca* but rather *Plurima ex paucissimis*—the most out of the least.

The Myth of Simplicity
Chapter 5
Section 4 (p. 82)

Bunting, Basil

I hate Science. It denies a man's responsibility for his own deeds, abolishes the brotherhood that springs from God's fatherhood. It is a hectoring, dictating expertise, which makes the least lovable of the Church Fathers seem liberal by contrast. It is far easier for a Hitler or a Stalin to find a mock-scientific excuse for persecution than it was for Dominic to find a mock-Christian one.

Quoted in Victoria Forde
The Poetry of Basil Bunting
Chapter 6
Letter of January 1, 1947 to Louis Zukofsky (p. 156)

Burbank, Luther

There is not personal salvation; there is no national salvation, except through science.

Attributed

Burhoe, R.W.

. . . in the usual sense of a *science* is a discipline possessed of an empirically validated theoretical structure which can indeed explain or account for and not simply describe, categorize, and correlate patterns of human experience/behavior.

Zygon
The Source of Civilization in the Natural Selection of Coadapted Information in Genes and Culture (p. 264)
Volume 11, Number 3, September 1976

Burroughs, John

Science has made or is making the world over for us. It has builded us a new house—builded it over our heads while we were yet living in the old, and the confusion and disruption and the wiping-out of the old features and the old associations, have been, and still are, a sore trial—a much finer, more spacious and commodious house . . . but new, new, all bright and hard and unfamiliar . . .

The Atlantic Monthly
In the Noon of Science (p. 327)
Volume 110, Number 3, September 1912

Bush, Vannevar

Science does not exclude faith . . . science does not teach a harsh materialism. It does not teach anything beyond its boundaries, and those boundaries have been severely limited by science itself.

Modern Arms and Free Men
Threat and Bulwark (p. 183)

Science has a simple faith, which transcends utility. Nearly all men of science, all men of learning for that matter, and men of simple ways too, have it in some form and in some degree. It is the faith that it is the privilege of man to learn to understand, and that this is his mission. If we abandon that mission under stress we shall abandon it forever, for stress will not cease. Knowledge for the sake of understanding, not merely to prevail, that is the essence of our being. None can define its limits, or set its ultimate boundaries.

Science Is Not Enough
The Search for Understanding (p. 191)

Butler, Samuel

Science is being daily more and more personified and anthropomorphized into a god. By and by they will say that science took our nature upon him, and sent down his only begotten son, Charles Darwin, or Huxley, into the world so that those who believe in him, etc.; and they will burn people for saying that science, after all, is only an expression for our ignorance of our own ignorance.

In Geoffrey Keynes and Brian Hill (Editors)
Samuel Butler's Notebooks
Science (p. 233)

Campbell, Norman R.

. . . science [is] the study of those judgments concerning which universal agreement can be obtained.

What Is Science?
Chapter II (p. 32)

Campbell, Thomas

Oh! star-eyed Science, hast thou wandered there,
To waft us home the message of dispair?

Poetical Works
Pleasures of Hope
Part II, L. 325

When Science from Creation's face
Enchantment's veil withdraws,
What lovely visions yield their place
To cold material laws!

Poetical Works
To the Rainbow
L. 13–16

Carlyle, Thomas

. . . what we might call by way of eminence, the *dismal science*.

The Nigger Question (p. 8)

Respectable Professors of the Dismal Science.

Latter Day Pamphlets
Number 1 (p. 38)

Chandrasekhar, S.

The pursuit of science has often been compared to the scaling of mountains, high and not so high. But who amongst us can hope, even in imagination, to scale the Everest and reach its summit when the sky is blue and the air is still, and in the stillness of the air survey the entire Himalayan range in the dazzling white of the snow stretching to infinity? None of us can hope for a comparable vision of nature and of the universe around us. But there is nothing mean or lowly in standing in the valley below and awaiting the sun to rise over Kinchinjunga.

Truth and Beauty
Chapter Two, X (pp. 26–7)

Chargaff, Erwin

What counts, however, in science is to be not so much the first as the last.

Science
Preface to a Grammar of Biology (p. 639)
Volume 172, Number 3984, May 1971

In science we always know much less than we believe we do.

Chemical and Engineering News
Uncertainties Great, Is the Gain Worth the Risk?
May 30, 1977

Chernin, Kim

Science is not neutral in its judgments, not dispassionate, nor detached . . .

The Obsession
Chapter 3 (p. 37)

Chesterton, G.K.

Science in the modern world has many uses; its chief use, however, is to provide long words to cover the errors of the rich.

Heretics
Celts and Celtophiles (p. 171)

. . . modern science cares far less for pure logic than a dancing Dervish.

Orthodoxy
The Maniac (p. 36)

When once one believes in a creed, one is proud of its complexity, as scientists are proud in the complexity of science. It shows how right it is in its discoveries. If it is right at all, it is a compliment to say that it is elaborately right. A stick might fit a hole or a stone a hollow by accident. But a key and a lock are both complex. And if a key fits a lock, you know it is the right key.

Orthodoxy
The Paradoxes of Christianity (p. 151)

[Modern science moves] toward the supernatural with the rapidity of a railway train.

Orthodoxy
Authority and the Adventurer (p. 277)

Science finds facts in Nature, but Science is not Nature; because Science has co-ordinated ideas, interpretations and analyses; and can say of Nature what Nature cannot say for itself.

The Resurrection of Rome
Chapter IV (p. 126)

Science, that nameless being, declared that the weakest must go to the wall; especially in Wall Street.

The Well and the Shallows
The Return to Religion (p. 74)

. . . physical science is like simple addition: it is either infallible or it is false.

All Things Considered
Science and Religion (p. 187)

Churchill, Winston

I have seldom seen a precise demand made upon science by the military which has not been met.

In F.B. Czarnomski
The Wisdom of Winston Churchill (p. 327)
Speech
Commons
March 16, 1950

It is arguable whether the human race have been gainers by the march of science beyond the steam engine. Electricity opens a field of infinite conveniences to ever greater numbers, but they may well have to pay dearly for them. But anyhow in my thought I stop short of the internal combustion engine which has made the world so much smaller. Still more must we fear the consequences of entrusting to a human race so little different from their predecessors of the so-called barbarous ages such awful agencies as the atomic bomb. Give me the horse.

In F.B. Czarnomski
The Wisdom of Winston Churchill (p. 327)
Speech
Royal College of Physicians
July 10, 1951

My experience—and it is somewhat considerable—is that in these matters when the need is clearly explained by military and political authorities, science is always able to provide something. "Seek and ye shall find" has been borne out.

In F.B. Czarnomski
The Wisdom of Winston Churchill (p. 327)
Speech
Commons
June 7, 1935

In the fires of science, burning with increasing heat every year, all the most dearly loved conventions are being melted down; and this is a process which is going continually to spread. In view of the inventions and discoveries which are being made for us, one might almost say every month, a unified direction of the war efforts of the three services would be highly beneficial.

In F.B. Czarnomski
The Wisdom of Winston Churchill (p. 327)
Speech
Commons
March 21, 1934

The latest refinements of science are linked with the cruelties of the Stone Age.

Speech
March 26, 1942
Source unknown

Science has given to this generation the means of unlimited disaster or of unlimited progress. There will remain the greater task of directing knowledge lastingly towards the purpose of peace and human good.

Speech
New Delhi, January 3, 1944
Source unknown

Science bestowed immense new powers on man and at the same time created conditions which were largely beyond his comprehension and still more beyond his control.

Speech
March 31, 1949
Source unknown

Cohen, Morris R.
The certainty which science aims to bring about is not a psychologic feeling about a given proposition but a logical ground on which its claim to truth can be founded.

Reason and Nature
Chapter Three, Section II(A) (p. 84)

Coles, Abraham
I value science—none can prize it more—
It gives ten thousand motives to adore.
Be it religious, as it ought to be,
The heart it humbles, and it bows the knee . . .

The Microcosm and Other Poems
The Microcosm
Christian Science

Collingwood, R.G.
A man ceases to be a beginner in any given science and becomes a master in that science when he has learned that . . . he is going to be a beginner all his life.

The New Leviathan
Part 1, Chapter 1, aphorism I.46

Comfort, Alex

Rash speculation does not bother the physicists—it has got them where they are today. And it is high time that the life sciences looked critically at the solidity of their tribal idols, including stochastic–genetic evolution, morphogenesis and the "mind–body problem"—while being mindful that, in the present climate, work on some quite unrelated matter may prove, incidentally and quite unwittingly, to have altered the entire face of the problem. Nor will the answers obtained lie within any existing frame of discourse.

Perspectives in Biology and Medicine
On Physics and Biology: Getting Our Act Together (p. 9)
Volume 29, Number 1, Autumn 1985

Commoner, Barry

Science is triumphant with far-ranging success, but its triumph is somehow clouded by growing difficulties in providing for the simple necessities of human life on the earth.

Science and Survival
Chapter 2 (p. 9)

Compton, Karl Taylor

Fundamentally, science means simply knowledge of our environment. Combined with ingenuity, science becomes power.

A Scientist Speaks (p. 2)

There is "something new under the sun" in that modern science has given mankind, for the first time in the history of the human race, a way of securing a more abundant life which does not simply consist in taking it away from someone else. Science really creates wealth and opportunity where they did not exist before.

A Scientist Speaks (p. 2)

I believe that the advent of modern science is the most important social event in all history.

A Scientist Speaks (p. 2)

Conant, James Bryant

There is only one proved method of assisting the advancement of pure science—that of picking men of genius, backing them heavily, and leaving them to direct themselves.

New York Times
Letter
August 13, 1945

[Science is] the activity of people who work in laboratories and whose discoveries have made possible modern industry and medicine.

Science and Common Sense
Chapter Two (p. 23)

Science is an interconnected series of concepts and schemes that have developed as a result of experimentation and observation and are fruitful of further experimentation and observation.

Science and Common Sense
Chapter Two (p. 25)

Condorcet, Jean

In every century Princes have been found to love the sciences and even to cultivate them, to attract Savants to their palaces and to reward by their favors and their amity men who afforded them a sure and constant refuge from world-weariness, a sort of disease to which supreme power seems particularly prone.

Éloge des académiciens de l'Académie royale des sciences
Forward, I

Constitution of the United States

The Congress shall have the Power . . . to promote the Progress of Science and useful Arts . . .

Article I, Section 8

Cooper, Leon

To say that science is logical is like saying that a painting is paint . . .

In George Johnson
In the Palaces of Memory
The Memory Machine (p. 194)

Crick, Francis

It can be confidently stated that our present knowledge of the brain is so primitive—approximately at the stage of the four humours in medicine or of bleeding in therapy (what is psychoanalysis but mental bleeding?)—that when we do have fuller knowledge our whole picture of ourselves is bound to change radically. Much that is now culturally acceptable will then seem to be nonsense. People with training in the arts still feel that in spite of the alterations made in their life by technology—by the internal combustion engine, by penicillin, by the Bomb—modern science has little to do with what concerns them most deeply. As far as today's science is concerned this is partly true, but tomorrow's science is going to knock their culture right out from under them.

Of Molecules and Men (p. 94–5)

Crothers, Samuel McChord

Science will not tolerate half knowledge nor pleasant imaginings, nor sympathetic appreciations; it must have definite demonstrations. The knowledge of the best that has been said and thought may be consoling, but it implies an unscientific principle of selection. It can be proved by statistics that the best things are exceptional.

The Gentle Reader
The Hinter-Land of Science (p. 228)

On the coasts of the Dark Continent of Ignorance the several sciences have gained a foothold. In each case there is a well-defined country carefully surveyed and guarded. Within its frontiers the laws are obeyed, and all affairs are carried on in an orderly fashion. Beyond it is a vague "sphere of influence," a Hinter-land over which ambitious claims of suzerainty are made; but the native tribes have not yet been exterminated, and life goes on very much as in the olden time.

The Gentle Reader
The Hinter-Land of Science (p. 231)

Cudmore, L.L. Larison

Almost anyone can do science; almost no one can do good science.

The Center of Life
Biochemical Evolution (p. 35)

. . . good science is almost always so very simple. *After* it has been done by someone else, of course.

The Center of Life
Biochemical Evolution (p. 36)

Curie, Eve

What does it matter to Science if her passionate servants are rich or poor, happy or unhappy, healthy or ill? She knows that they have been created to seek and to discover, and that they will seek and find until their strength dries up at its source. It is not in a scientist's power to struggle against his vocation: even on his days of disgust or rebellion his steps lead him inevitably back to his laboratory apparatus.

Madame Curie
Chapter XV (p. 193)

Curie, Marie

In science we must be interested in things, not in persons.

In Eve Curie
Madame Curie
Chapter XVI (p. 222)

da Vinci, Leonardo

Science is the captain, practice the soldiers.

Leonardo da Vinci's Notebooks (p. 54)

All sciences which end in words are dead the moment they come to life, except for their manual part, that is to say, the writing, which is the mechanical part.

The Literary Works of Leonardo da Vinci
Volume I, 7a1148 (p. 35)

Dante Alighieri
. . . no Science demonstrates its own subject, but presupposes it.

In G. Giuliani
Il Convito di Dante Alighieri
Tract II, Cap. XIV (p. 142)

Darwin, Sir Francis
But in science the credit goes to the man who convinces the world, not to the man to whom the idea first occurs.

Eugenics Review
First Galton Lecture before the Eugenics Society
Volume 6, Number 1, 1914

Forgive me for suggesting one caution; as Demosthenes said, "Action, action, action," was the soul of eloquence, so is caution almost the soul of science.

More Letters of Charles Darwin
Volume II
To Dohrn
January 4, 1870 (p. 444)

How grand is the onward rush of science; it is enough to console us for the many errors which we have committed, and for our efforts being overlaid and forgotten in the mass of new facts and new views which are daily turning up.

The Life and Letters of Charles Darwin
Volume II
Darwin to Wallace
August 28, 1872 (p. 348)

de Gourmont, Remy
Science is the only truth and it is the great lie. It knows nothing, and people think it knows everything. It is misrepresented. People think that science is electricity, automobilism, and dirigible balloons. It is something very different. It is life devouring itself. It is the sensibility transformed into intelligence. It is the need to know stifling the need to live. It is the genius of knowledge vivisecting the vital genius.

In Glen S. Burne (Editor/Translator)
Selected Writings
Part II
Art and Science (p. 172)

Descartes, René

Science in its entirety is true and evident cognition. He is no more learned who has doubts on many matters than the man who has never thought of them; nay he appears to be less learned if he has formed wrong opinions on any particulars. Hence it were better not to study at all than to occupy one's self with objects of such difficulty, that, owing to our inability to distinguish true from false, we are forced to regard the doubtful as certain; for in those matters any hope of augmenting our knowledge is exceeded by the risk of diminishing it. Thus in accordance with the above maxim we reject all such merely probable knowledge and make it a rule to trust only what is completely known and incapable of being doubted.

Rules for the Direction of the Mind
Rule II

Dewey, John

The routine of custom tends to deaden even scientific inquiry; it stands in the way of *discovery* and of the *active* scientific worker. For discovery and inquiry are synonymous as an occupation. Science is a *pursuit*, not a coming into possession of the immutable; new theories as points of view are more prized than discoveries that quantitatively increase the store on hand.

Reconstruction in Philosophy
Introduction (p. xvii)

Dickinson, Emily

I'll climb the "Hill of Science,"
I "view the landscape o'er,"
Such transcendental prospects,
I ne'er beheld before!

The Complete Poems of Emily Dickinson
#3

"Arcturus" is his other name—
I'd rather call him "Star."
It's very mean of Science
To go and interfere!

The Complete Poems of Emily Dickinson
#70

Dickinson, G. Lowes

Science hangs in a void of nescience, a planet turning in the dark.

A Modern Symposium (p. 159)

Disraeli, Benjamin

What Art was to the ancient world, Science is to the modern . . .

Coningsby
Book iv, Chapter 1

. . . the pursuit of science leads only to the insoluble.

Lothair
Chapter XVII (p. 70)

Dobie, J. Frank

Putting on the spectacles of science in expectation of finding the answer to everything looked at signifies inner blindness.

The Voice of the Coyote
Introduction (p. xvi)

Drake, Daniel

The love of pleasure and the love of science may coexist, but cannot be indulged at the same time; though in fact they are seldom united. A student should draw his pleasures from the discovery of truth, and find his amusements in the beauties and wonders of nature. He should seek for recreation not debauchery. The former invigorates the mind, the latter enervates it. Study and recreation, properly alternated, bring out the glorious results of rich and powerful thought, original conceptions, and elevated design; dissipation wastes the whole, perverts the moral taste and impoverishes the intellect. One makes great men—the other wild men. One creates the sun—the other the comet of the social and scientific heavens. A fixed luminox, spending life and light on all around us, is one—a wandering flashing and vanishing meteor, is the other.

Introductory Lectures, on the Means of Promoting the Intellectual Improvement of Students (p. 5)

Dryden, John

Science distinguishes a Man of Honour from one of those Athletick Brutes whom undeservedly we call Heros.

Fables Ancient and Modern
Dedication

Is it not evident, in these last hundred years (when the Study of Philosophy has been the business of all the *Virtuosi* in *Christendome*) that almost a new Nature has been reveal'd to us? that more errours of the School have been detected, more useful Experiments in Philosophy have been made, more Noble Secrets in Opticks, Medicine, Anatomy, Astronomy, discover'd, than in all those credulous and doting Ages from *Aristotle* to us? so true it is that nothing spreads more fast than Science, when rightly and generally cultivated.

Of Dramatick Poesie: An Essay
Volume I (p. 12)

du Noüy, Pierre Lecomte

Either we have absolute confidence in our science and in the mathematical and other reasonings which enable us to give a satisfactory explanation of the phenomena surrounding us—in which case we are forced to recognize that certain fundamental problems escape us and that their explanation amounts to admitting a miracle—or else we doubt the universality of our science and the possibility of explaining all natural phenomena by chance alone; and we fall back on a miracle or a hyperscientific intervention.

Human Destiny
Chapter 3 (p. 36)

Science is very young when compared to the moral, spiritual, and religious ideas of humanity. It enjoys the prestige of a new toy. But we must not be misled. In spite of its youth and imperfections, it constitutes the best means of convincing us of the immensity and harmonious beauty of the universe, revealed by the infinite complexity of the apparently most simple phenomena. It is an orderly and confounding complexity which is a thousand times better qualified than ignorance to make us feel the omnipotence of the Creator.

Between Knowing and Believing
The Future of Spirit 1941 (p. 216)

Dubos, René

Science is still the versatile, unpredictable hero of the play, creating endless new situations, opening romantic vistas and challenging accepted concepts.

Louis Pasteur, Free Lance of Science
Chapter I (p. 15)

Science knows no country, because knowledge belongs to humanity, and is the torch which illuminates the world. Science is the highest personification of the nation because that nation will remain the first which carries the furthest the works of thought and intelligence.

Louis Pasteur, Free Lance of Science
Chapter III (p. 85)

Science is not the product of lofty meditations and genteel behavior, it is fertilized by heartbreaking toil and long vigils—even if, only too often, those who harvest the fruit are but the laborers of the eleventh hour.

Louis Pasteur, Free Lance of Science
Chapter XIV (p. 389)

Dyson, Freeman

If we try to squeeze science into a single philosophical viewpoint like reductionism we are like Procrustes chopping off the feet of his guests when they do not fit on his bed.

In J. Cornwell (Editor)
Nature's Imagination? The Frontiers of Scientific Vision
The Scientist as Rebel (p. 11)

Eben, Aubry

Science is not a sacred cow. Science is a horse. Don't worship it. Feed it.

Quoted in *Reader's Digest*
March 1963

Eddington, Sir Arthur Stanley

To imagine that Newton's great scientific reputation is tossing up and down in these latter-day revolutions is to confuse science with omniscience.

The Nature of the Physical World
Chapter X (p. 202)

Einstein, Albert

The grand aim of all science . . . is to cover the greatest number of empirical facts by logical deductions from the smallest number of hypotheses or axioms.

Life
In Lincoln Barnett
The Meaning of Einstein's New Theory (p. 22)
January 9, 1950

. . . science can only ascertain what *is*, but not what *should be*, and outside of its domain value judgments of all kinds remain necessary.

Out of My Later Years
Science and Religion
II (p. 25)

The whole of science is nothing more than a refinement of everyday thinking.

Out of My Later Years
Physics and Reality
I (p. 59)

The aim of science is, on the one hand, a comprehension, as *complete* as possible, of the connection between the sense experiences in their totality, and, on the other hand, the accomplishment of this aim *by the use of a minimum of primary concepts and relations*.

Out of My Later Years
Physics and Reality
I (p. 63)

Science as something existing and complete is the most objective thing known to man. But science in the making, science as an end to be pursued, is as subjective and psychologically conditioned as any other branch of human endeavor—so much so, that the question "what is the purpose and meaning of science?" receives quite different answers at different times and from different sorts of people . . .

The World as I See It
Address at Columbia University, New York (p. 137)

Science is a wonderful thing if one does not have to earn one's living at it.

Letter to a California student March 24, 1951
Quoted in Helen Dukas and Banesh Hoffmann
Albert Einstein: The Human Side (p. 57)

Science is the attempt to make the chaotic diversity of our sense experience correspond to a logically uniform system of thought.

Science
Considerations Concerning the Fundaments of Theoretical Physics (p. 487)
Volume 91, Number 2369, May 24, 1940

Strange that science, which in the old days seemed harmless, should have evolved into a nightmare that causes everyone to tremble.

In G.J. Whitrow
Einstein: The Man and His Achievement
Chapter III (p. 89)

Science will stagnate if it is made to serve practical goals.

In Otto Nathan and Heinz Norden
Einstein on Peace
Chapter Thirteen (p. 402)

One thing I have learned in a long life: that all our science, measured against reality, is primitive and childlike—and yet it is the most precious thing we have.

In Banesh Hoffmann
Albert Einstein: Creator and Rebel (p. v)

Science is the century-old endeavor to bring together by means of systematic thought the perceptible phenomena of this world into as thoroughgoing an association as possible.

In Conference on Science, Philosophy and Religion
Science, Philosophy and Religion
Chapter XIII
Science and Religion (p. 209)

Einstein, Albert
Infeld, Leopold

Science is not just a collection of laws, a catalogue of unrelated facts. It is a creation of the human mind, with its freely invented ideas and concepts.

The Evolution of Physics
Quanta (p. 310)

Eiseley, Loren

In the end, science as we know it has two basic types of practitioners. One is the educated man who still has a controlled sense of wonder before the universal mystery, whether it hides in a snail's eye or within the light that impinges on that delicate organ. The second kind of observer is the extreme reductionist who is so busy stripping things apart that the tremendous mystery has been reduced to a trifle, to intangibles not worth troubling one's head about.

The Star Thrower
Science and the Sense of the Holy (p. 190)

Eisenschiml, Otto

Science seeks to build, not to destroy; to aid, not to hinder.

The Art of Worldly Wisdom
Part I (p. 8)

Emelyanov, A.S.

Science provides mankind with a great tool of cognition which makes it possible to reach unprecedented heights of abundance and equality. This determines the most important and most fruitful aspect of the social role of science, and as a result the social responsibility of scientists is growing. A scientist cannot be a 'pure' mathematician, biophysicist or sociologist for he cannot remain indifferent to the fruits of his work, to whether they will be useful or harmful to mankind. An indifferent attitude as to whether people will be better or worse off as a result of scientific achievement is cynicism, if not a crime.

Quoted by E.H.S. Burhop
In Maurice Goldsmith and Alan Mackay (Editors)
Society and Science
Scientist and Public Affairs (p. 31)

Emerson, Ralph Waldo

Men love to wonder, and that is the seed of our science.

Society and Solitude
Works and Days (p. 142)

Feigl, H.
. . . science, properly interpreted, is not dependent on any sort of metaphysics. It merely attempts to cover a maximum of facts by a minimum of laws.

American Quarterly
Naturalism *and* Humanism (p. 148)
Volume 1, Number 2, Summer 1949

Ferré, Nels F.S.
It is a sad experience to hear someone denounce science as the cause of modern chaos and destruction. Our technological advance may be abused and make of what could be a near heaven a near hell, but that is surely not the fault of science as such. Science has not failed man, but man has failed science.

Faith and Reason
Chapter II (p. 38)

Science is supposed by many to have banished every realm of the sacred; and behold, science becomes the sacred cow.

Faith and Reason
Chapter II (p. 43)

Feyerabend, Paul
Science is an essentially anarchistic enterprise . . .

Against Method
Analytical Index (p. 10)

Fischer, Emil
It is no use, Gentlemen, science is and remains international.

In Albert Einstein
The World as I See It
The International Science (p. 50)

Fiske, John
. . . all human science is but the increment of the power of the eye . . .

The Destiny of Man
VII (p. 60)

Flaubert, Gustave
My kingdom is as wide as the world, and my desire has no limit. I go forward always, freeing spirits and weighing worlds, without fear, without compassion, without love, and without God. Men call me science.

La Tentation de Saint-Antoine

Freud, Sigmund

Science in her perpetual incompleteness and insufficiency is driven to hope for her salvation in new discoveries and new ways of regarding things. She does well, in order not to be deceived, to arm herself with skepticism and to accept nothing new unless it has withstood the strictest examination.

In J.A.V. Butler
Science and Human Life
Chapter Fifteen (p. 120)

No, our science is no illusion. But an illusion it would be to suppose that what science cannot give us we can get elsewhere.

The Future of an Illusion
Chapter X (p. 92)

Friend, J.W.
Feibleman, James K.

The modern world, which has lost faith in so many causes, still accepts science nearly unchallenged. Science to-day occupies the position held by the Roman Church in the Middle Ages: as the single great authority in a world divided on almost every object of loyalty.

What Science Really Means
Chapter I (p. 11)

Fromm, Erich

The pace of science forces the pace of the technique. Theoretical physics forces atomic energy on us; the successful production of the fission bomb forces upon us the manufacture of the hydrogen bomb. *We* do not choose our problems, we do not choose our products; we are pushed, we are forced—by what? By a system which has no purpose and goal transcending it, and which makes man its appendix.

The Sane Society
Chapter Five
Nineteenth-Century Capitalism (p. 83)

Galbraith, John Kenneth

The real accomplishment of modern science and technology consists in taking ordinary men, informing them narrowly and deeply and then, through appropriate organization, arranging to have their knowledge combined with that of other specialized but equally ordinary men. This dispenses with the need for genius. The resulting performance, though less inspiring, is far more predictable.

The New Industrial State
Chapter VI (p. 73)

Galton, Francis

A special taste for science seems frequently to be so ingrained in the constitution of scientific men, that it asserts itself throughout their whole existence.

In Karl Pearson
The Life, Letters and Labours of Francis Galton
Volume II (p. 152)

Glass, H. Bentley

. . . the general citizen of his country, the man in the street, must learn what science is, not just what it can bring about. Surely this is our primary task. If we fail in this, then within a brief period of years we may expect either nuclear devastation or worldwide tyranny. It is not safe for apes to play with atoms. Neither can men who have relinquished their birthright of scientific knowledge expect to rule themselves.

In Hilary J. Deason
A Guide to Science Reading
Revolution in Biology (pp. 25–6)

Goethe, Johann Wolfgang von

When a science appears to be slowing down and, despite the efforts of many energetic individuals, comes to a dead stop, the fault is often to be found in a certain basic concept that treats the subject too conventionally. Or the fault may lie in a terminology which, once introduced, is unconditionally approved and adopted by the great majority, and which is discarded with reluctance even by independent thinkers, and only as individuals in isolated cases.

Botanical Writings
The Metamorphosis of Plants
An Attempt to Evolve a General Comparative Theory (p. 81)

Four epochs of science:
childlike,
poetic, superstitious;
empirical,
searching, curious;
dogmatic,
didactic, pedantic;
ideal,
methodical, mystical.

Scientific Studies
Volume 12
Chapter VIII (pp. 304–5)

Sciences destroy themselves in two ways: by the breadth they reach and by the depth they plumb.

Scientific Studies
Volume 12
Chapter VIII (p. 305)

A crisis must necessarily arise when a field of knowledge matures enough to become a science, for those who focus on details and treat them as separate will be set against those who have their eye on the universal and try to fit the particular into it.

Scientific Studies
Volume 12
Chapter VIII (p. 305)

Germans—and they are not alone in this—have a knack of making the sciences unapproachable.

Scientific Studies
Volume 12
Chapter VIII (p. 306)

In general the sciences put some distance between themselves and life, and make their way back to it only by a roundabout path.

Scientific Studies
Volume 12
Chapter VIII (p. 306)

The present age has a bad habit of being abstruse in the sciences. We remove ourselves from common sense without opening up a higher one; we become transcendent, fantastic, fearful of intuitive perception in the real world, and when we wish to enter the practical realm, or need to, we suddenly turn atomistic and mechanical.

Scientific Studies
Volume 12
Chapter VIII (pp. 308–9)

Goldstein, A.

Science is always a race . . . and scientists are competitive people. Because the monetary rewards are minimal, they go for ego rewards . . .

Quoted in J. Goldberg
Anatomy of a Scientific Discovery
Locks and Keys (p. 25)

Gornick, Vivian

To do science today is to experience a dimension unique in contemporary working lives; the work promises something incomparable: the sense of living both personally and historically. That is why science now draws to itself all kinds of people—charlatans, mediocrities, geniuses—everyone who wants to touch the flame, feel alive in the time.

Women in Science
Part One (p. 26)

Gould, Laurence M.

Today, there is no other influence comparable with science in changing the foundations, indeed the very character of our lives. Science and its products determine our economy, dominate our industry, affect our health and welfare, alter our relations to all other nations, and determine the conditions of war and peace. Everyone who breathes is affected, and cannot remain impervious to them.

UNESCO Courier
Science and the Culture of Our Times (p. 4)
February 1968

Gould, Stephen Jay

Science, since people must do it, is a socially embedded activity. It progresses by hunch, vision, and intuition. Much of its change through time does not record a closer approach to absolute truth, but the alteration of cultural contexts that influence it so strongly. Facts are not pure and unsullied bits of information; culture also influences what we see and how we see it. Theories, moreover, are not inexorable inductions from facts. The most creative theories are often imaginative visions imposed upon facts; the source of imagination is also strongly cultural.

The Mismeasure of Man
Introduction (pp. 21–2)

Gray, Thomas

While bright-eyed Science watches round . . .

The Complete Poetical Works of Gray, Beattie, Blair, Collins, Thomson, and Kirke White
Elegy Written in a Country Churchyard
Ode for Music
Ode vii
I, L. 11

Here rests his head upon the lap of Earth,
A youth to fortune and to fame unknown.
Fair Science frown'd not on his humble birth,
And Melancholy mark'd him for her own.

The Complete Poetical Works of Gray, Beattie, Blair, Collins, Thomson, and Kirke White
Elegy Written in a Country Churchyard
The Epitaph
Stanza 1

Grove, Sir William

For my part I must say that science to me generally ceases to be interesting as it becomes useful.

In H.B.G. Casimir
Haphazard Reality
Chapter 8 (p. 226)

Gruenberg, Benjamin C.

To vast numbers of men and women science appears as something altogether too remote from their interests or capacities to justify even a glance, or a hope of grasping. It is something for the "highbrows" or wizards.

Science and the Public Mind
Chapter XIV (p. 152)

Handler, Philip
My own belief is that science remains the most powerful tool we have yet generated to apply leverage for our future. It is the instrument which is most useful for guiding our own destinies, for assuring the condition of man in the years to come. I have much to hope that we will not abandon that tool, leaving us to our own brute devices.

Hearings, 1971 National Science Foundation Authorization
Subcommittee on Science, Research and Development
House Committee on Science and Astronautics
91st Congress, 2nd session 1970 (p. 16)

Harari, Josué V.
Bell, David F.
Science is the totality of the world's legends. The world is the space of their inscription. To read and to journey are one and the same act.

In Michel Serres
Hermes: Literature, Science, Philosophy
Introduction (p. xxi)

Hardy, Godfrey
A science is said to be 'useful' if its development increases, even indirectly, the material well-being and comfort of men, if it promotes happiness, using that word in a crude and commonplace way.

A Mathematician's Apology
Section 19 (pp. 115–6)

Harman, Willis
. . . since we have come to the understanding that science is not a description of "reality" but a metaphorical ordering of experience, the new science does not impugn the old. It is not a question of which view is "true" in some ultimate sense. Rather, it is a matter of which picture is more useful in guiding human affairs.

In Philip R. Lee *et al* (Editors)
Symposium on Consciousness
Quoted by Philip R. Lee and Frances Petrocelli
Can Consciousness Make a Difference? (p. 3)

Harrington, John
Science is the progressive discovery of the nature of nature.

Dance of the Continents
The Lure of the Hunt (p. 30)

Harris, Errole E.

Accordingly there are two main types of science, exact science . . . and empirical science . . . seeking laws which are generalizations from particular experiences and are verifiable (or, more strictly, 'probabilifiable') only by observation and experiment.

Hypothesis and Perception
Prevalent Views of Science (p. 25)

Harrison, Jane

Science has given us back something strangely like a World-Soul . . .

Ancient Art and Ritual
Chapter VII (p. 246)

Harvey, William

Although there is but one road to science, that, to wit, in which we proceed from things more known to things less known, from matters more manifest to matters more obscure; and universals are principally known to us, science bringing by reasoning from universals to particulars; still the comprehension of universals by the understanding is based upon the perception of individual things by the senses.

Anatomical Exercises on the Generation of Animals
Of the Manner and Order of Acquiring Knowledge (p. 332)

Haskins, C.P.

Science provides a challenge to effort for the individual youth that is far greater than the challenge of militarism. It provides a unity of thought which is far wider, for it transcends all national boundaries. It provides a wider battleground, for the goal of militarism is the conquering of man, but that of science is the understanding and the subjugation of all the rest of our natural environment. And finally, it is an infinitely broader training than totalitarian training can possibly be, for it requires, in addition to great courage, stamina, and drive, the qualities of intellectualism and gentleness.

In Conference on Science, Philosophy and Religion
Science, Philosophy and Religion
Scientific Thought and a Democratic Ideology (p. 235)

Havel, Václav

Modern science . . . abolishes as mere fiction the innermost foundations of our natural world: it kills God and takes his place on the vacant throne so henceforth it would be science that would hold the order of being in its hand as its sole legitimate guardian and so be the legitimate arbiter of all relevant truth . . . People thought they could explain and conquer nature—yet the outcome is that they destroyed it and disinherited themselves from it.

In Lewis Wolpert
The Unnatural Nature of Science
Introduction (p. ix)

Hazlitt, William

The origin of all science is in the desire to know causes; and the origin of all false science and imposture is in the desire to accept false causes rather than none; or, which is the same thing, in the unwillingness to acknowledge our own ignorance.

The Atlas
February 15, 1829
Burke and the Edinburgh Phrenologists
This article is unsigned in the atlas but appears in P.P. Howe's
New Writings by William Hazlitt, 1925

Heisenberg, Werner

Natural science does not simply describe and explain nature; it is part of the interplay between nature and ourselves; it describes nature as exposed to our method of questioning.

Physics and Philosophy
Chapter V (p. 81)

In science, too, it is impossible to open up new territory unless one is prepared to leave the safe anchorage of established doctrine and run the risk of a hazardous leap forward. With his relativity theory, Einstein had abandoned the concept of simultaneity, which was part of the solid ground of traditional physics, and, in so doing, outraged many leading physicists and philosophers and turned them into bitter opponents. In general, scientific progress calls for no more than the absorption and elaboration of new ideas—and this is a call most scientists are happy to heed.

Physics and Beyond
Chapter 6 (p. 70)

Henderson, Lawrence

Science has finally put the old teleology to death. Its dismembered spirit, freed from vitalism and all material ties, immortal, alone lives on, and from such a ghost science has nothing to fear.

The Fitness of the Environment
Chapter VIII
Section III, B (p. 311)

Henry, William

In every science it is necessary clearly to distinguish between what is certain and what is merely probable; the laws of combination, in definite and multiple proportions, appear to me to belong to the former class; but the generalization, which explains those truths by speculations respecting atoms, must be acknowledged, in its present state, to be entirely theoretical.

Elements of Experimental Chemistry
11th edition
I (p. 53)

Herschel, John Frederick William

Science is the knowledge of many, orderly and methodically digested and arranged, so as to become attainable by one. The knowledge of reasons and their conclusions constitutes *abstract*, that of causes and their effects, and of the laws of nature, *natural science*.

A Preliminary Discourse on the Study of Natural Philosophy
Chapter II, Section 13 (p. 18)

Hertz, Heinrich

The rigour of science requires that we distinguish well the undraped figure of nature itself from the gay-coloured vesture with which we clothe it at our pleasure.

Quoted by Ludwig Boltzmann
Letter
Nature
February 28, 1895

Herzen, Alexander

Science, which cuts its way through the muddy pond of daily life without mingling with it, casts its wealth to right and left, but the puny boatmen do not know how to fish for it.

My Past and Thoughts
The Later Years
Fragments
Swiss Views (p. 592)

Superficial dilettantism and the narrow specialization of the scientists *ex officio* are the two banks of science which prevent the fertilizing waters of this Nile from overflowing.

Selected Philosophical Works
Dilettantism in Science (p. 52)

Science is a table abundantly laid for every man whose hunger is great enough, whose craving for spiritual nourishment has grown sufficiently insistent.

Selected Philosophical Works
Dilettantism in Science (p. 58)

Science, in the best sense of the word, shall come to be accessible to the people, and when it does it shall claim a voice in all practical matters.

Selected Philosophical Works
Dilettantism in Science (p. 69)

Highet, Gilbert
There are naive people all over the world—some of them scientists—who believe that all problems, sooner or later, will be solved by Science. The word Science itself has become a vague reassuring noise, with a very ill-defined meaning and a powerful emotional charge: It is now applied to all sorts of unsuitable subjects and used as a cover for careless and incomplete thinking in dozens of fields. But even taking Science at the most sensible of its definitions, we must acknowledge that it is unperfect as are all activities of the human mind.

Man's Unconquerable Mind
Chapter 4 (p. 106)

Hilbert, David
Our Science, which we loved above everything, had brought us together. It appeared to us as a flowering garden. In this garden there were well-worn paths where one might look around at leisure and enjoy oneself without effort, especially at the side of a congenial companion. But we also liked to seek out hidden trails and discovered many an unexpected view which was pleasing to our eyes; and when the one pointed it out to the other, and we admired it together, our joy was complete.

Memorial address for Hermann Minkowski
In S. Chandrasekhar
Truth and Beauty
Chapter Three (p. 52)

Hill, D.W.
Science, then, is more than technology and is greater than the sciences. It is a system of thought, a philosophy and a guide to maturity. It is a living thing of joy and beauty intimately interwoven with the affairs of life and yet distinct from them. It is a medium of expression in which the imagination has full play. In short, the method of science has resulted for the first time not only the beauties and harmonies of the universe but also the ingenuity of the human mind which can build such a superstructure as the natural sciences on the foundation of observable facts.

Science
Chapter I (pp. 8–9)

Hilts, Philip
In all human activities, it is not ideas of machines that dominate; it is people. I have heard people speak of "the effect of personality on science." But this is a backward thought. Rather, we should talk about the

effect of science on personalities. Science is not the dispassionate analysis of impartial data. It is the human, and thus passionate, exercise of skill and sense on such data. Science is not an exercise in which objectivity is prized.

Scientific Temperaments: Three Lives in Contemporary Science
Preface (pp. 11–2)

Hocking, R.

It is an oversimplification to compare the impersonal aspect of science with the impersonal aspects of industrial society, and to deplore both in one breath. The former is an achievement of self forgetful concentration upon truths about nature. The latter are deplorable to the extent that they exhibit crude power of men over men. By contrast, the selflessness of the scientific calling does silent honor to personal existence.

In T.J.J. Altizer, William A. Beardslee, and J. Harvey Young (Editors)
Truth, Myth, and Symbol
The Problem of Truth (p. 5)

Holmes, Oliver Wendell

Science is the topography of ignorance.

The Writings of Oliver Wendell Holmes
Volume IX
Border Lines in Medical Science (p. 211)

Go on, fair science; soon to thee
  Shall nature yield her idle boast;
Her vulgar fingers formed a tree,
  But thou hast trained it to a post.

The Complete Poetical Works of Oliver Wendell Holmes
The Meeting of the Dryads (p. 412)

I am a little afraid that science is breeding us down too fast into coral-insects. A man like Newton or Leibnitz or Haller used to paint a picture of outward or inward nature with a free hand, and stand back and look at it as a whole and feel like an archangel; but nowadays you have a Society, and they come together and make a great mosaic, each man bringing his little bit and sticking it in its place, but so taken up with his petty fragment that he never thinks of looking at the picture the little bits make when they are put together.

The Poet at the Breakfast-Table (p. 92)

Horgan, J.

. . . to pursue science in a speculative, postempirical mode that I call ironic science. Ironic science resembles literary criticism in that it offers points of view, opinions, which are, at best, interesting, which provoke further comment. But it does not converge on the truth. It cannot achieve empirically verifiable surprises that force scientists to make substantial revisions in their basic description of reality.

The End of Science
Introduction (p. 7)

Hubbard, Elbert
SCIENCE: 1. The knowledge of the common people classified and carried one step further. 2. Accurate organized knowledge grounded on fact. 3. Classified superstition.

The Roycroft Dictionary (p. 134)

Hubbard, Ruth
To overturn orthodoxy is no easier in science than in philosophy, religion, economics, or any of the other disciplines through which we try to comprehend the world and the society in which we live.

Women Look at Biology Looking at Women
Have Only Men Evolved? (p. 10)

Hubble, Edwin
. . . equipped with his five senses, man explores the universe around him and calls the adventure science.

The Nature of Science
The Nature of Science (p. 6)

Huizinga, Johan
. . . that all science is merely a game can be easily discarded as a piece of wisdom too easily come by. But it is legitimate to enquire whether science is not liable to indulge in play within the closed precincts of its own method. Thus, for instance, the scientist's continuous penchant for systems tends in the direction of play.

Homo Ludens
Chapter XII (p. 203)

Huxley, Aldous
For Science in its totality, the ultimate goal is the creation of a monistic system in which—on the symbolic level and in terms of the inferred components of invisibility and intangibly fine structure—the world's enormous multiplicity is reduced to something like unity, and the endless successions of unique events of a great many different kinds get tidied and simplified into a single rational order. Whether this goal will ever be reached remains to be seen. Meanwhile we have the various sciences, each with its own system coordinating concepts, its own criterion of explanation.

Literature and Science
Chapter 3 (p. 9)

Science is a matter of disinterested observation, patient ratiocination within some system of logically correlated concepts. In real-life conflicts between reason and passion the issue is uncertain. Passion and prejudice are always able to mobilize their forces more rapidly and press the attack with greater fury; but in the long run (and often, of course, too late) enlightened self-interest may rouse itself, launch a counterattack and win the day for reason.

Literature and Science
Chapter 23 (p. 68)

Science sometimes builds new bridges between universes of discourse and experience hitherto regarded as separate and heterogeneous. But science also breaks down old bridges and opens gulfs between universes that, traditionally, had been connected.

Literature and Science
Chapter 37 (p. 111)

To the twentieth century man of letters science offers a treasure of newly discovered facts and tentative hypotheses. If he accepts this gift and if, above all, he is sufficiently talented and resourceful to be able to transform the new materials into works of literary art, the twentieth century man of letters will be able to treat the age-old, and perennially relevant, theme of human destiny with a depth of understanding, a width of reference of which, before the rise of science, his predecessors (through no fault of their own, no defect of genius) were incapable.

Literature and Science
Chapter 29 (p. 87)

All science is based upon an act of faith—faith in the validity of the mind's logical processes, faith in the ultimate explicability of the world, faith that the laws of thought are laws of things.

Ends and Means
Beliefs (p. 298)

. . . science has 'explained' nothing; the more we know the more fantastic the world becomes and the profounder the surrounding darkness.

Along the Road
Part II
Views of Holland (p. 108)

Huxley, Julian

Sciences, like Empires, have their rise and their time of flourishing, though not their decay.

What Dare I Think?
Chapter I (p. 1)

Science has two main functions in civilization. One is to give man a picture of the world phenomena, the most accurate and complete picture possible. The other is to provide him with the means of controlling his environment and his destiny.

What Dare I Think?
Chapter IV (pp. 127–8)

Without the impersonal guidance and the efficient control provided by science civilization will either stagnate or collapse, and human nature cannot make progress towards realizing its possible evolutionary destiny.

What Dare I Think?
Chapter V (p. 177)

Science . . . has not only turned her face outwards from man, but stripped him of all the robes of his divinity, turned him out of the palace that he had so laboriously built in the center of the world, and left him in rags, pitiably insignificant and suddenly transported to an outlying corner of the cosmos.

Harper's Monthly Magazine
Will Science Destroy Religion? (p. 535)
April 1926

The attempt to understand this universe, including the nature of man, is the task of science; and as she makes progress with this task, so will she become more and more an indispensable part of philosophy and religion—imagination's touchstone, thought's background, action's base.

The Century Illustrated Monthly Magazine
Searching for the Elixir of Life
Volume 103, Number 4, February 1922

Huxley, Thomas
Nothing great in science has ever been done by men, whatever their powers, in whom the divine afflatus of the truth-seeker was wanting.

Collected Essays
Volume I
Methods and Results
The Progress of Science (p. 56)

. . . whatever evil voices may rage, Science, secure among the powers that are eternal, will do her work and be blessed.

Collected Essays
Volume I
Methods and Results
Descartes' Discourse on Method (p. 198)

Extinguished theologians lie about the cradle of every science as the strangled snakes beside that of Hercules . . .

Collected Essays
Volume II
Darwiniana
The Origin of Species (p. 52)

. . . science . . . commits suicide when it adopts a creed.

Collected Essays
Volume II
Darwiniana
The Darwin Memorial (p. 252)

So, the vast results obtained by Science are won by no mystical faculties, by no mental processes, other than those which are practised by every one of us, in the humblest and meanest affairs of life. A detective policeman discovers a burglar from the marks made by his shoe, by a mental process identical with that by which Cuvier restored the extinct animals of Montmartre from fragments of their bones.

Collected Essays
Volume III
Science and Education
On the Educational Value of the Natural History Sciences (p. 46)

Whatever happens, science may bide her time in patience and in confidence.

Collected Essays
Volume V
Science and Christian Tradition
An Episcopal Trilogy (p. 143)

The generalizations of science sweep on in ever-widening circles, and more aspiring flights, through limitless creation.

Letter to *The Times*
December 26, 1859

What men of science want is only a fair day's wages for more than a fair day's work, . . .

Method and Results
Essays
Administrative Nihilism (p. 287)

Addressing myself to you, as teachers, I would say, mere book learning in physical science is a sham and a delusion—what you teach, unless you wish to be impostors, that you must first know; and real knowledge in science means personal acquaintance with the facts, be they few or many.

Collected Essays
Volume VIII
The Study of Zoology (p. 227)

Ingersoll, Robert G.
Reason, Observation, and Experience—the Holy Trinity of Science.

On the Gods and Other Essays
The Gods (p. 54)

Jacks, L.P.
Science is never static, never stagnant, never content with the boundary it has reached. It is always dynamic, always breaking bounds . . . Science . . . abhors a limitation.

The Atlantic Monthly
Is There a Foolproof Science? (p. 231)
February 1924

. . . in limiting science, we are also limiting ourselves.

The Atlantic Monthly
Is There a Foolproof Science? (p. 233)
February 1924

Science is the pursuer, life is the pursued . . .

The Atlantic Monthly
Is There a Foolproof Science? (p. 238)
February 1924

Jacob, François
For science, there are many possible worlds; but the interesting one is the world that exists and has already shown itself to be at work for a long time. Science attempts to confront the possible with the actual.

The Possible and the Actual
Myth and Science (p. 12)

James, William
Science herself consults her heart when she lays it down that the infinite ascertainment of fact and correction of false belief are the supreme gods for man.

The Will to Believe and Other Essays in Popular Philosophy
The Will to Believe (p. 22)

Science as such assuredly has no authority, for she can only say what is, not what is not.

The Will to Believe
Is Life Worth Living? (p. 56)

Science like life feeds on its own decay. New facts burst old rules; then newly developed concepts bind old and new together into a reconciling law.

The Will to Believe
Psychical Research (p. 320)

The aim of 'Science' is to attain conceptions so adequate and exact that we shall never need to change them.

Principles of Psychology
Volume 2
The Perception of Things
Apperception (p. 109)

Jevons, William Stanley

Science arises from the discovery of Identity amidst Diversity.

The Principles of Science
Chapter I (p. 1)

Jones, Steve

This is the essence of science. Even though I do not understand quantum mechanics or the nerve cell membrane, I trust those who do. Most scientists are quite ignorant about most sciences but all use a shared grammar that allows them to recognize their craft when they see it. The motto of the Royal Society of London is 'Nullius in verba': trust not in words. Observation and experiment are what count, not opinion and introspection. Few working scientists have much respect for those who try to interpret nature in metaphysical terms. For most wearers of white coats, philosophy is to science as pornography is to sex: it is cheaper, easier, and some people seem, bafflingly, to prefer it. Outside of psychology it plays almost no part in the functions of the research machine.

The New York Review of Books
Review of *How the Mind Works* by Steve Pinker (p. 13)
November 6, 1997

Joubert, Joseph

How many scholars there are who hammer out knowledge; hardworking, enthusiastic, indefatigable Cyclopes—with only one eye.

Pensées
XVII (p. 119)

Kaczynski, Theodore

Science marches on blindly . . . without regard to the real welfare of the human race or to any other standard, obedient only to the psychological needs of the scientists and of the government officials and corporate executives who provide the funds for research.

Scientific American
In Anne Eisenberg
The Unabomber and the Bland Decade (p. 35)
Volume 274, Number 4, April 1998

Kapitza, Peter Leonidovich

The year that Rutherford died there disappeared for ever the happy days of free scientific work which gave us such delight in our youth. Science has lost her freedom. Science has become a productive force. She has become rich but she has become enslaved and part of her is veiled in secrecy.

Nature
Address to the Royal Society in Honour of Lord Rutherford (p. 783)
Volume 210, May 17, 1966

Kellog, Vernon

Science does not assume that it knows—despite the great deal that it does know—more than a very small part of the order of nature.

The World's Work
Some Things Science Doesn't Know (p. 528)
March 1926

Kennedy, John F.

Let both sides seek to invoke the wonders of science instead of its terrors. Together let us explore the stars, conquer the deserts, eradicate disease, tap the ocean depths and encourage the arts and commerce.

Inaugural Address
January 20, 1961

Science contributes to our culture in many ways, as a creative intellectual activity in its own right, as the light which has served to illuminate man's place in the universe, and as the source of understanding of man's own nature.

Address to the National Academy of Sciences
Washington, D.C.
October 22, 1963

Kettering, Charles F.

So that we might kill one another more expertly, science found wonderful ways to live more comfortably, richly, to communicate more rapidly. So that we might exterminate one another more successfully, science showed us how we might all live longer and stronger . . .

In Paul de Kruif
Saturday Evening Post
America Comes through a Crisis (p. 3)
May 13, 1933

Keyser, Cassius J.

Not in the ground of need, not in bent and painful toil, but in the deep-centered play-instinct of the world, in the joyous mood of the eternal Being, which is always young, Science has her origin and root; and her spirit, which is the spirit of genius in moments of elevation, is but a sublimated form of play, the austere and lofty analogue of the kitten playing with the entangled skein or of the eaglet sporting with the mountain winds.

Mathematics (p. 44)

King, Martin Luther, Jr.

We have genuflected before the god of science only to find that it has given us the atomic bomb, producing fears and anxieties that science can never mitigate.

Strength To Love
Chapter XIII (p. 106)

Kingsley, Charles

. . . Science was the child of Courage, and Courage the child of Knowledge.

Health and Education
Science (p. 259)

For from blind fear of the unknown, science does certainly deliver man. She does by man as he does by an unbroken colt. The colt sees by the road side some quite new object—a cast-away boot, an old kettle, or what not. What a fearful monster! What unknown terrific powers may it not possess! And the colt shies across the road, runs up the bank, rears on end; putting itself thereby, as many a man does, in real danger. What cure is there? But one, experience. So science takes us, as we should take the colt, gently by the halter; and makes us simply smell at the new monster; till after a few trembling sniffs, we discover, like the colt, that it is not a monster, but a kettle.

Health and Education
Science (p. 284)

For science is, I verily believe, like virtue, its own exceeding great reward.

Health and Education
Science (p. 289)

Kirkpatrick, Clifford

Science recognizes no personal powers in the universe responsive to the prayers and needs of men.

Religion in Human Affairs
Chapter XVI (p. 470)

Köhler, Wolfgang

. . . at the present time it is of course quite customary for physicists to trespass on chemical ground, for mathematicians to do excellent work in physics, and for physicists to develop new mathematical procedures . . . trespassing is one of the most successful techniques in science.

Dynamics in Psychology
Retention and Recall (p. 116)

Koshland, Daniel E., Jr.

Science. Science is not impressed with a conglomeration of data. It likes carefully constructed analysis of each problem.

Science
Editorial
January 14, 1994

Kroeber, A.L.

. . . it appears that the total work of science must be done on a series of levels which the experience of science gradually discovers.

The Nature of Culture (p. 121)

Krutch, Joseph Wood

Science has always promised two things not necessarily related—an increase first in our powers, second in our happiness or wisdom, and we have come to realize that it is the first and less important of the two promises which it has kept most abundantly.

The Modern Temper
The Disillusion with the Laboratory (p. 43)

. . . the most important part of our lives—our sensations, emotions, desires, and aspirations—takes place in a universe of illusions which science can attenuate or destroy, but which it is powerless to enrich.

The Modern Temper
The Disillusion with the Laboratory (p. 50)

Though many have tried, no one has ever yet explained away the decisive fact that science, which can do so much, cannot decide what it ought to do.

The Measure of the Man
The Loss of Confidence (p. 31)

Kuhn, Thomas S.

. . . science . . . often suppresses fundamental novelties because they are necessarily subversive of its basic commitments.

The Structure of Scientific Revolutions
Chapter I (p. 5)

Normal science . . . is predicated on the assumption that the scientific community knows what the world is like. Much of the success of the enterprise derives from the community's willingness to defend that assumption, if necessary at considerable cost.

The Structure of Scientific Revolutions
Chapter I (p. 5)

Lamb, Charles

Science has succeeded to poetry, no less in the little walks of children than with men. Is there no possibility of averting this sore evil?

Letter to S.T. Coleridge
October 23, 1802

Can we unlearn the arts that pretend to civilize, and then burn the world? There is a march of science; but who shall beat the drums for its retreat?

Letter to George Dyer
December 20, 1830

In everything that relates to *science*, I am a whole Encyclopaedia behind the rest of the world.

The Essays of Elia
The Old and the New Schoolmaster (p. 57)

Lapp, Ralph E.

No one—not even the most brilliant scientist alive today—really knows where science is taking us. We are aboard a train which is gathering speed, racing down a track on which are an unknown number of switches leading to unknown destinations. No single scientist is in the cab, and there may be demons at the switch. Most of society is in the caboose looking backward. Some passengers, fearful that they have boarded an express train to hell, want to jump off before it is too late. The opposition, it would appear, is no longer open, but at least the passengers can discuss matters among themselves, and keep a hand on the brake.

The New Priesthood
Chapter 2 (p. 29)

Larrabee, Eric

Science is a—what? a method, a faith, a body of facts, a structure of theories, an institution, a way of life, a finite number of duly qualified individuals, an infinity of relevance and possibility. For a large number of scientists, science is indescribable, but indisputably a *thing*: it is knowable, palpable, reliable, usable. They live with it and by it; it is simply and unequivocally *there*.

Commentary
Science and the Common Reader (p. 43)
June 1966

Laudan, Larry
The aim of science is merely to secure theories with a high problem-solving effectiveness.

New Scientist
August 1, 1892 (p. 26)

Le Guin, Ursula K.
. . . it is only when science asks why, instead of simply describing how, that it becomes more than technology. When it asks why, it discovers Relativity. When it only shows how, it invents the atomic bomb and then puts its hands over its eyes and says, *My God what have I done?* . . .

Language of the Night
The Stalin in the Soul (p. 219)

Lebowitz, Fran
Science is not a pretty thing. It is unpleasantly proportioned, outlandishly attired and often over-eager. What then is the appeal of science? What accounts for its popularity? And who gives it its start?

Metropolitan Life
Science (p. 76)

. . . modern science was largely conceived of as an answer to the servant problem and that it is generally practiced by those who lack the flair for conversation.

Metropolitan Life
Science (p. 76)

Lerner, Max
It is not science that has destroyed the world, despite all the gloomy forebodings of the earlier prophets. It is man who has destroyed man.

Actions and Passions
The Human Heart and Human Will (p. 3)

Lewis, Gilbert N.
The strength of science lies in its naïvete.

The Anatomy of Science
Chapter I (p. 1)

Lewis, Wyndham

When we say "science" we can either mean any manipulation of the inventive and organizing power of the human intellect: or we can mean such an extremely different thing as the *religion of science*, the vulgarized derivative from this pure activity manipulated by a sort of priestcraft into a great religious and political weapon.

The Art of Being Ruled
Revolution and Progress
Chapter 1 (pp. 3–4)

The puritanic potentialities of science have never been forecast. If it evolves a body of organized rites, and is established as a religion, hierarchically organized, things more than anything else will be done in the name of 'decency.' The course fumes of tobacco and liquors, the consequent tainting of the breath and staining of white fingers and teeth, which is so offensive to many women, will be the first things attended to.

The Art of Being Ruled
The Family and Feminism
Chapter 7 (p. 210)

Lichtenberg, Georg Christoph

The most heated defenders of a science, who can not endure the slightest sneer at it, are commonly those who have not made very much progress in it and are secretly aware of this defect.

Aphorisms
Notebook F, Aphorism 8

There is no greater impediment to progress in the sciences than the desire to see it take place too quickly.

Aphorisms
Notebook K, Aphorism 72

Lubbock, John

Science, our Fairy Godmother, will, unless we perversely reject her help, and refuse her gifts, so richly endow us, that fewer hours of labour will serve to supply us with the material necessaries of life, leaving us more time to ourselves, more leisure to enjoy all that makes life best worth living.

The Beauties of Nature
Introduction (p. 37)

Lundberg, G.A.

. . . no science tells us *what to do* with the knowledge that constitutes the science. Science only provides a car and a chauffeur for us. It does not directly, as science, tell us where to drive. The car and the chauffeur will take us into the ditch, over the precipice, against a stone wall, or into the highlands of age-long human aspirations with equal efficiency. If we agree as to where we want to go and tell the driver our goal, he should be able to take us there by one of a number of possible routes the costs and conditions of each of which the scientist should be able to explain to us.

Can Science Save Us?
Social Problems (p. 31)

Lynch, Gary

What you're really seeking are constraints . . . You're seeking things that box you in. That's what separates science from most other human endeavors. Religion is not something where people sit down and say, Well, *if* there were a god *then* . . . But science is a constant search for that, for those things that hem you in.

In George Johnson
In the Palaces of Memory
Mucking Around in the Wetware (p. 91)

What you really need to do the best science is a tremendous tolerance for ambiguity. *You have to be able to tolerate ambiguity*. Because we as creatures are set up for some reason to see cause-and-effect. And what you really wind up doing is tolerating the fact that you have all these assumptions and all these uncertainties, and living with them. And when you really go into a novel area, what do you have to guide you? The more novel it is, the fewer the constraints. For a human being that is a very uncomfortable feeling.

In George Johnson
In the Palaces of Memory
Mucking Around in the Wetware (pp. 91–2)

Mach, Ernst

Physical science does not pretend to be a *complete* view of the world; it simply claims that it is working toward such a complete view in the future.

The Science of Mechanics (p. 464)

The function of science, as we take it, is to replace experience. Thus, on the one hand, science must remain in the province of experience, but, on the other, must hasten beyond it, constantly expecting confirmation, constantly expecting the reverse. Where neither confirmation nor refutation is possible, science is not concerned . . .

The Science of Mechanics (p. 586)

Magendie, François

I am a mere street scavenger of science. With hook in hand and basket on my back, I go about the streets of science collecting whatever I find.

In René Dubos
Louis Pasteur, Free Lance of Science
Chapter XIII (p. 363)

March, Robert H.

Science is more than a mere attempt to describe nature as accurately as possible. Frequently the real message is well hidden, and a law that gives a poor approximation to nature has more significance than one which works fairly well but is poisoned at the root.

Physics for Poets
Chapter I (p. 21)

Marguerite of Valosis

Science conducts us, step by step,
Through the whole range of creation, until
we arrive, at length, at God.

Memoirs
Letter XII

Mathieu, Emile

As the domain of science broadens and expands it becomes more and more necessary to expound its principles with clearness and conciseness and to substitute for artificial processes, however skillful, the transformations that can be accounted for by the nature of the subject.

Course of Mathematical Physics
Preface

Maxwell, James Clerk

It was a great step in science when men became convinced that, in order to understand the nature of things, they must begin by asking, not whether a thing is good or bad, noxious or beneficial, but of what kind it is? and how much is there of it? Quality and Quantity were then first recognized as the primary features to be observed in scientific inquiry.

In William H. George
The Scientist in Action
British Association Address, 1870 (p. 15)

McCarthy, Mary

Modern neurosis began with the discoveries of Copernicus. Science made man feel small by showing him that the earth was not the center of the universe.

On the Contrary
Tyranny of the Orgasm

Mead, Margaret

. . . the negative cautions of science are never popular. If the experimentalist would not commit himself, the social philosopher, the preacher and the pedagogue tried the harder to give a short-cut answer.

Coming of Age in Samoa
Chapter 1 (p. 3)

Medawar, Sir Peter

. . . the factual burden of a science varies inversely with its degree of maturity. As a science advances, particular facts are comprehended within, and therefore in a sense annihilated by, general statements of steadily increasing explanatory powers and compass. In all sciences we are being progressively relieved of the burden of singular instances, the tyranny of the particular. We need no longer record the fall of every apple.

The Art of the Soluble
Two Conceptions of Science (p. 114)

Mercer, de la Rivière

Without the sciences man would rank below the beasts.

Quoted in L. Ducros
Les Encyclopédistes (p. 315)

Meyer, Agnes

From the nineteenth century view of science as a god, the twentieth century has begun to see it as a devil. It behooves us now to understand that science is neither the one nor the other.

Education for a New Morality
Chapter 2 (p. 11)

Millikan, Robert A.

We need science in education, and much more of it than we now have, not primarily to train technicians for the industries, which demand them, though that may be important, but much more to give everybody a little glimpse of the scientific mode of approach to life's problems, to give everyone some familiarity with at least one field in which the distinction between right and wrong is not always blurred and uncertain, to let him see that it is not true that "one opinion is as good as another," . . .

Science
The Relationship of Science to Industry (p. 30)
Volume LXIX, Number 1776, January 11, 1929

Milne, E.A.

The Christmas message—which is also the Christian message—is *'Gloria in excelsis Deo'* . . . Glory to God in the highest and on earth peace among men of goodwill . . . This is not a bad definition of the aim of all true science: the aim of rejoicing in the splendid mysteries of the world and universe we live in, and of attempting so to understand those mysteries that we can improve our command over nature, improve our conditions of life and so ensure peace . . .

Modern Cosmology and the Christian Idea of God
Chapter I (p. 1)

Monod, Jacques

In science, self-satisfaction is death. Personal self-satisfaction is the death of the scientist. Collective self-satisfaction is the death of the research. It is restlessness, anxiety, dissatisfaction, agony of mind that nourish science.

News Science
Obituary (p. 359)
Volume 109, June 5, 1976

Montagu, Ashley

As the god of contemporary man's idolatry, science is a two-handed engine, and as such science is too important a human activity to leave to the scientist.

Quoted in
New York Times Book Review
April 26, 1964
Advertisement of Jacques Barzun's
Science: The Glorious Entertainment

More, Louis Trenchard

Science has so many dazzling achievements to its credit; we have done so many things which seemed to be impossible, that the popular mind is apt to conclude that, if an explanation is given in the name of science, it must be true whether it be understood or not.

The Dogma of Evolution
Chapter Seven (p. 241)

Moscovici, S.

Science has become involved in this adventure, our adventure, in order to renew everything it touches and warm all that it penetrates—the earth on which we live and the truths which enable us to live. At each turn it is not the echo of a demise, a bell tolling for a passing away that is heard, but the voice of rebirth and beginning, ever afresh, of mankind and materiality, fixed for an instant in their ephemeral permanence. That is why the great discoveries are not revealed on a deathbed like that of Copernicus, but offered like Kepler's on the road of dreams and passion.

Hommes domestiques et hommes sauvages
(pp. 297–8)

Movie
The benefits of science are not for scientists, Marie. They're for humanity.

The Story of Louis Pasteur
Paul Muni preaching in bed to his wife

Muller, Herbert J.
Although science is no doubt the Jehovah of the modern world, there is considerable doubt about the glory of its handiwork.

Science and Criticism
Chapter III (p. 59)

Mumford, Lewis
. . . however far modern science and technics have fallen short of their inherent possibilities, they have taught mankind at least one lesson: Nothing is impossible.

Technics and Civilization
Chapter VIII, Section 13 (p. 435)

O'Neill, Eugene
Darrell: Happiness hates the timid! So does Science!

Strange Interlude
Act Four (p. 152)

Orr, Louis
Science will never be able to reduce the value of a sunset to arithmetic. Nor can it reduce friendship to a formula. Laughter and love, pain and loneliness, the challenge of accomplishment in living, and the depth of insight into beauty and truth: these will always surpass the scientific mastery of nature.

Commencement address at Emory University, Atlanta
June 6, 1960

Osler, Sir William
. . . who can doubt that the leaven of science, working in the individual, leavens in some slight degree the whole social fabric. Reason is at least free, or nearly so; the shackles of dogma have been removed, and faith herself, freed from a morganatic alliance, finds in the release great gain.

Aequanimitas
The Leaven of Science (p. 94)

Science is organized knowledge, and knowledge is of things we see. Now the things that are seen are temporal; of the things that are unseen science knows nothing and has at present no means of knowing anything.

Science and Immortality
The Terestans (pp. 40–1)

The future belongs to science. More and more she will control the destinies of the nations. Already she has them in her crucible and on her balances.

In Harvey Cushing
The Life of Sir William Osler
Volume II (p. 262)

Pallister, William

You are the sum of what we know,
You are our might and main;
You are the whole of what is so,
The little we retain:
Our fond beliefs all come and go,
And you alone remain.

Poems of Science
Science (p. 39)

Science works by the slow method of the classification of data, arranging the detail patiently in a periodic system into groups of facts, in series like the strata of the rocks. For each series there must be a vocabulary of special words which do not always make good sense when used in another series. But the laws of periodicity seem to hold throughout, among the elements and in every sphere of thought, and we must learn to co-ordinate the whole through our new conception of the reign of relativity.

Poems of Science
Man and the Stars (p. 88)

Panunzio, Constantine

Science . . . involves active, purposeful search; it discovers, accumulates, sifts, orders, and tests data; it is a slow, painstaking, laborious activity; it is a search after bodies of knowledge sufficiently comprehensive to lead to the discovery of uniformities, sequential orders or so-called "laws"; it may be carried on by an individual, but it gains relevance only as it produces data which can be added to and tested by the findings of others.

Major Social Institutions
Chapter 20 (p. 322)

If science is to subserve human needs, it will continue to discover and catalogue "all the islands of the universe 300,000,000 or more light years distant," but it will not fiddle while Rome burns . . .

Major Social Institutions
Chapter 21 (p. 338)

Parton, H.N.

A successful blending of the sciences and the humanities is necessary for the health of our civilization.

Science is Human
Science is Human (p. 31)

Pasteur, Louis

I am imbued with two deep impressions; the first, that science knows no country; the second, which seems to contradict the first, although it is in reality a direct consequence of it, that science is the highest personification of the nation. Science knows no country, because knowledge belongs to humanity, and is the torch which illuminates the world, Science is the highest personification of the nation because that nation will remain the first which carries the furthest the works of thought and intelligence.

In René Dubos
Pasteur and Modern Science
Chapter 15
A Dedicated Life (p. 146)

Pavlov, Ivan P.

Only science, exact science about human nature itself, and the most sincere approach to it by the aid of the omnipotent scientific method, will deliver man from his present gloom, and will purge him from his contemporary shame in the sphere of interhuman relations.

Lectures on Conditioned Reflexes
Preface to the First Russian Edition (p. 41)

Pearson, Karl

Modern Science, as training the mind to an exact and impartial analysis of facts, is an education specifically fitted to promote sound citizenship.

The Grammar of Science
Introductory
Section 3 (p. 13)

When every fact, every present or past phenomenon of that universe, every phase or present or past life therein, has been examined, classified, and co-ordinated with the rest, then the mission of science will be completed. What is this but saying that the task of science can never end till man ceases to be, till history is no longer made, and development itself ceases?

The Grammar of Science
Introductory
Section 5 (p. 16)

Every great advance of science opens our eyes to facts which we have failed before to observe, and makes new demands on our powers of interpretation. This extension of the material of science into regions where our great-grandfathers could see nothing at all, or where they would have declared human knowledge impossible, is one of the most remarkable features of modern progress. Where they interpreted the motion of the planets of our own system, we discuss the chemical constitution of stars, many of which did not exist for them, for the telescopes could not reach them. Where they discovered the circulation of the blood, we see the physical conflict of living poisons within the blood, whose battles would have been absurdities for them.

The Grammar of Science
Introductory
Section 5 (p. 17–8)

Does science leave no mystery? On the contrary it proclaims mystery where others profess knowledge. There is mystery enough in the universe of sensation and in its capacity for containing those little corners of consciousness which project their own products, of order and law and reason, into an unknown and unknowable world. There is mystery enough here, only let us clearly distinguish it from ignorance within the field of possible knowledge. The one is impenetrable, the other we are daily subduing.

The Grammar of Science
The Scientific Law
Conclusion (p. 97)

Science for the past is a description, for the future a belief . . .

The Grammar of Science
Cause and Effect—Probability
Section 1 (p. 99)

Peattie, Donald Culross

Of our windows on the universe, science is set with the clearest pane; it is not warped or waved to make the images appear to support any dogma; the glass is not rose-tinted, neither is it leaded with a picture that shuts out the sun and, coming between the light of day and you, enforces the credence of the past upon the young present.

Flowering Earth
Chapter 18 (p. 244)

Planck, Max

That we do not construct the external world to suit our own ends in the pursuit of science, but that *vice versa* the external world forces itself upon our recognition with its own elemental power, is a point which ought to be categorically asserted again and again in these positivistic times. From the fact that in studying the happenings of nature we strive to eliminate the contingent and accidental and to come finally to what is essential and necessary, it is clear that we always look for the basic thing behind the dependent thing, for what is absolute behind what is relative, for the reality behind the appearance and for what abides behind what is transitory. In my opinion, this is characteristic not only of physical science but all of science.

Where Is Science Going?
From Relative to Absolute (pp. 198–9)

Science cannot solve the ultimate mystery of nature. And that is because, in the last analysis, we ourselves are part of nature and therefore part of the mystery that we are trying to solve.

Where Is Science Going?
Epilogue (p. 217)

Poe, Edgar Allen
Science! true daughter of old Time thou art
Who alterest all things with thy peering eyes!
Why prey'st thou thus upon the poet's heart,
Vulture! whose wings are dull realities!

Al Aaraaf, Tamerlane and Minor Poems (p. 11)

Poincaré, Henri
Man, then cannot be happy through science, but to-day he can be much less be happy without it.

The Foundations of Science
The Value of Science
Introduction (p. 206)

The advance of science is not comparable to the changes of a city, where old edifices are pitilessly torn down to give place to new, but to the continuous evolution of zoologic types which develop ceaselessly and end by becoming unrecognizable to the common sight, but where an expert eye finds always traces of the prior work of the past centuries.

The Foundations of Science
The Value of Science
Introduction (p. 208)

Polanyi, M.
The morsels of science which [the young scientist] picks up—even though often dry or else speciously varnished—instill in him the intimation of intellectual treasures and creative joys far beyond his ken. His intuitive realization of a great system of valid thought and of an endless path of discovery sustain him in laboriously accumulating knowledge and urge him on to penetrate into intricate brain-racking theories. Sometimes he will also find a master whose work he admires and whose manner and outlook he accepts for his guidance. Thus his mind will become

assimilated to the premise of science. The scientific institution of reality henceforth shapes his perception. He learns the methods of scientific investigation and accepts the standards of scientific value.

Science, Faith and Society (p. 44)

This coherence of valuation throughout the whole range of science underlies the unity of science. It means that any statement recognized as valid in one part of science can, in general, be considered as underwritten by all scientists. It also results in a general homogeneity of and a mutual respect between all kinds of scientists, by virtue of which science forms an organic unity.

Science, Faith and Society (p. 49)

Pope, Alexander

One science only will one genius fit,
So vast is art, so narrow human wit;

The Works of Alexander Pope
Volume II
Essay on Criticism
Part I, L. 60–61

Go, wondrous creature! mount where science guides
Go, measure earth, weigh air, and state the tides;
Instruct the planets in what orbs to run,
Correct old Time, and regulate the sun;

. . .

Go teach Eternal Wisdom how to rule—
Then drop into thyself, and be a fool!

The Works of Alexander Pope
Volume II
Essay on Man
Epistle II, L. 19–22, 29–30

How Index-learning turns no student pale,
Yet holds the eel of science by the tail: . . .

The Works of Alexander Pope
Volume IV
Dunciad
Book I, L. 279–280

Far eastward cast thine eye, from whence the Sun
And orient Science their brite course begun: . . .

The Works of Alexander Pope
Volume IV
Dunciad
Book III, L. 73–74

Beneath her footstools, Science groans in Chains, . . .

The Works of Alexander Pope
Volume IV
Dunciad
Book IV, L. 21

Pope Paul VI
It is all too evident that science is not self-sufficient, cannot be an end in itself. Science is made by man for man; it must therefore come out of the sphere of its research and reflect on man and through man on human society and universal history . . . The faithful scientist cannot, confronted with the consequents of his discovery, ignore the complex being that is, in the last analysis, the human person.

Quote
August 7, 1966

Popper, Karl R.
Science does not aim, primarily, at high probabilities. It aims at a high informative content, well backed by experience. But a hypothesis may be very probable simply because it tells us nothing, or very little.

The Logic of Scientific Discovery
Appendix ix (p. 399)

. . . what is to be called 'science' and who is to be called a 'scientist' must always remain a matter of convention or decision.

The Logic of Scientific Discovery
Chapter II, Section 10 (p. 52)

The empirical basis of objective science has thus nothing "absolute" about it. Science does not rest upon solid bedrock. The bold structure of its theories rises, as it were, above a swamp. It is like a building erected on piles. The piles are driven down from above into the swamp, but not down to any natural or "given" base; and when we cease our attempts to drive our piles into a deeper layer, it is not because we have reached firm ground. We simply stop when we are satisfied that they are firm enough to carry the structure, at least for the time being.

The Logic of Scientific Discovery
Chapter V, Section 30 (p. 111)

Science is not a system of certain, or well-established, statements; nor is it a system which steadily advances towards a state of finality . . . *We do not know: we can only guess*. And our guesses are guided by the unscientific . . . faith in laws, in regularities which we can uncover–discover.

The Logic of Scientific Discovery
Chapter X, Section 85 (p. 278)

Science may be described as the art of systematic over-simplification.

Quoted in
The Observer
London, August 1, 1982

. . . it is the aim of science to find *satisfactory explanations*, of whatever strikes us as being in need of explanation.

Objective Knowledge: An Evolutionary Approach
Chapter 5 (p. 191)

Porter, Sir George

Should we force science down the throats of those that have no taste for it? Is it our duty to drag them kicking and screaming into the twenty-first century? I am afraid that it is.

Source unknown
Speech 1966

Porterfield, Austin L.

Science, in the broadest sense, is the entire body of the most accurately tested, critically established, systematized knowledge available about that part of the universe which has come under human observation. For the most part this knowledge concerns the forces impinging upon human beings in the serious business of living and thus affecting man's adjustment to and of the physical and the social world . . . Pure science is more interested in understanding, and applied science is more interested in control . . .

Creative Factors in Scientific Research
Chapter II (p. 11)

Poteat, William Louis

Science confers power, not purpose. It is a blessing, therefore, if the purpose which it serves is good; it is a curse, if the purpose is bad.

Can a Man Be a Christian Today?
Part I, Section 2 (p. 27)

Praed, Winthrop

Of science and logic he chatters,
  As fine and as fast as he can;
Though I am no judge of such matters,
  I'm sure he's a talented man.

The Poems of Winthrop Mackworth Praed
The Talented Man

Pratt, C.C.

Science is a vast and impressive tautology.

The Logic of Modern Psychology
Chapter VI (p. 154)

Prescott, William Hickling

It is the characteristic of true science, to discern the impassable, but not very obvious, limits which divide the province of reason from that of speculation. Such knowledge comes tardily. How many ages have rolled away in which powers, that, rightly directed, might have revealed the great laws of nature, have been wasted in brilliant, but barren reveries on alchemy and astrology.

History of the Conquest of Mexico
Volume I
Book I, Chapter IV (p. 132)

Prior, Matthew

Forc'd by reflective Reason I confess,
That human Science is uncertain guess.

Matthew Prior's Literary Works
Volume I
Solomon
Book 1, L. 740–741

Pritchett, V.S.

A touch of science, even bogus science, gives an edge to the superstitious tale.

The Living Novel & Later Appreciations
An Irish Ghost (p. 123)

Proverb

The sciences have bitter roots, but their fruits are sweet.

In Robert Christy
Proverbs, Maxims and Phrases of All Ages (p. 236)

Proverb, Arabic

Science is a plant whose roots indeed are at Mecca, but its fruit ripens at Herat.

In Robert Christy
Proverbs, Maxims and Phrases of All Ages (p. 236)

Proverb, Spanish

Ciencia es locura,
Si buen sentido no la cura.
[Science is madness
If good sense does not cure it.]

Source unknown

Quine, W.V.O.
Science is like a boat, which we rebuild plank by plank while staying afloat in it. The philosopher and the scientist are in the same boat.

In George Johnson
The Palaces of Memory
The End of Philosophy (p. 222)

Ratzinger, Cardinal Joseph
Science is not an absolute to which all things have to be subordinated.

Observer
Sayings of the Week
March 15, 1987

Ravetz, J.R.
The obsolescence of the conception of science as the pursuit of truth results from several changes in the social activity of science. First, the heavy warfare with "theology and metaphysics" is over. Although a few sharp skirmishes still occur, the attacks on the freedom of science from this quarter are no longer significant. This is not so much because of the undoubted victory of science over its ancient contenders as for the deeper reason that the conclusions of natural science are no longer ideologically sensitive. What people, either the masses or the educated, believe about the inanimate universe or the biological aspects of humanity is not relevant to the stability of society as it was once thought to be.

Scientific Knowledge and Its Social Problems
Chapter I (pp. 20–1)

Renan, Ernest
Science, and science alone, can give to humanity what it most craves, a symbol and a law.

Quoted in Louis Trenchard More
The Limitations of Science
Chapter I (p. 2)

A little true science is better than a great deal of bad science. One is less liable to error by confessing one's ignorance than by fancying that one knows a great many things one does not.

The Future of Science
Preface (p. xix)

. . . science must pursue its road without minding with whom it comes in collision. Let the others get out of the way. If it appears to raise objections against received dogmas, it is not for science but the received dogmas to be on the defensive and to reply to the objections. Science should behave as if the world were free from preconceived opinions, and not heed the difficulties it starts.

The Future of Science
Chapter V (p. 83)

. . . science is a religion, science alone will henceforth make the creeds, science alone can solve for men the eternal problems, the solutions of which his nature imperatively demands.

The Future of Science
Chapter V (p. 97)

The lofty serenity of science becomes possible only on the condition of impartial criticism, which without regard for the beliefs of a certain portion of humanity, handles its imperturbable instrument with the inflexibility of the geometrician, without anger and without pity. The critic never insults.

The Future of Science
Chapter XV (p. 257)

Richards, I.A.

For science, which is simply our most elaborate way of *pointing* to things systematically, tells us and can tell us nothing about the nature of things in any *ultimate* sense. It can never answer any question of the form: *What* is so and so? It can only tell us *how* so and so behaves. And it does not attempt to do more than this.

Science and Poetry
Chapter V (pp. 52–3)

Richet, Charles

All . . . believe in the sovereignty of science; which like the grammar of Martine, rules even over kings, and imperiously subjects them to its laws.

The Natural History of a Savant
Chapter II (p. 13)

No one has the right to encumber science with premature assertions.

The Natural History of a Savant
Chapter X (p. 122)

To neglect science is to exclude fair hope, to condemn ourselves to live in a uniform monotonous existence.

The Natural History of a Savant
Chapter XIII (p. 147)

The future and the happiness of humanity depend on science.

The Natural History of a Savant
Chapter XIII (p. 155)

Romanoff, Alexis

Science is advanced by husbands, but wives are often behind them.

Encyclopedia of Thoughts
Aphorism 91

Into the life of a cultured man enter science, art, and poetic philosophy.

Encyclopedia of Thoughts
Aphorism 1220

Pure science has no part in politics.

Encyclopedia of Thoughts
Aphorism 1281

Dedication to pure science or fine arts often is incompatible with a desire for economic gain.

Encyclopedia of Thoughts
Aphorism 1457

Religions offer unbounded faith; science, logical preciseness; and the arts, creative imagination.

Encyclopedia of Thoughts
Aphorism 1471

Science and technology may lead to self-destruction; humanities to sensible social recovery.

Encyclopedia of Thoughts
Aphorism 2937

Rubin, Harry

. . . one of the great pitfalls of science is the fallacy of misplaced concreteness. Scientists seem to prefer questionable explanations to no explanation at all.

Journal of the National Cancer Institute
Does Somatic Mutation Cause Most Cancers? (p. 999)
Volume 64, Number 5, May 1980

Ruse, Michael

Science, like most human cultural phenomena, has evolved. What was allowable in the early nineteenth century is not necessarily allowable in the late twentieth century. Specifically, science today does not break with law. And this is what counts for us. We want criteria of science for today, not for yesterday.

Science, Technology & Human Values
Response to the Commentary: *Pro Judice* (p. 21)
Volume 7, Number 41, Fall 1982

Ruskin, John

Science does its duty, not in telling us the causes of spots in the sun, but in explaining to us the laws of our own life, and the consequences of their violation.

In Henry Attwell
Thoughts From Ruskin (p. 29)

Russell, Bertrand

In science men have discovered an activity of the very highest value in which they are no longer, as in art, dependent for progress upon the appearance of continually greater genius, for in science the successors stand upon the shoulders of their predecessors; where one man of supreme genius has invented a method, a thousand lesser men can apply it.

A Free Man's Worship and Other Essays
Chapter 3

Sagan, Carl

One of the great commandments of science is, "Mistrust arguments from authority."

The Demon-Haunted World
Chapter 2 (p. 28)

Sand, Ole

Science is not the be-all and end-all of life. We must keep our debt to it in clear perspective. Its penicillin has saved us, its wash-and-wear has clothed us; its air-conditioning has cooled us. One day, its promises of moon-living may even give us the universe. But the test tube has yet to come up with an easy formula for increasing man's ability to think, to feel, to appreciate. It is the task of the humanities to help us understand ourselves, as well as our fellow men, and to help us live in this brave new world that science has fashioned for the Seventies.

Music Educators Journal
Schools for the Seventies
July 17, 1966

Sandage, Allan

Science is the only self-correcting human institution, but it also is a process that progresses only by showing itself to be wrong.

In Alan Lightman and Roberta Brawer
Origins
Allan Sandage (p. 82)

Santayana, George

Science is a half-way house between private sensation and universal vision.

The Life of Reason
Part V, Chapter I (p. 385)

Schiller, F.C.S.

Science: To one, she is the exhalted and heavenly Goddess; to another she is a capable cow which keeps him supplied with butter.

In Folke Dovring
Knowledge and Ignorance (p. 141)

To Archimedes once came a youth, who for knowledge was thirsting,
Saying, "Initiate me into the science divine,
Which for my country has borne forth fruit of such wonderful value,
And which the wall of the town 'gainst the Sambuca protects.
"Calls't thou the science divine? It *is* so," the wise man responded;
"But it was so, my son, ere it avail'd for the town.
Wouldst thou have fruit from her only, e'en mortals wit that can
provide thee;
Wouldst thou the goddess obtain, seek not the woman in Her!"

Schiller's Poems
Archimedes and the Student

Schrödinger, Erwin

. . . there is a tendency to forget that all science is bound up with human culture in general, and that scientific findings, even those which at the moment appear the most advanced and esoteric and difficult to grasp, are meaningless outside their cultural context. A theoretical science unaware that those of its constructs considered relevant and momentous are destined eventually to be framed in concepts and words that have a grip on the educated community and become part and parcel of the general world picture—a theoretical science, I say, where this is forgotten, and where the initiated continue musing to each other in terms that are, at best, understood by a small group of close fellow travelers, will necessarily be cut off from the rest of cultural mankind; in the long run it is bound to atrophy and ossify however virulently esoteric chat may continue within its joyfully isolated groups of experts.

The British Journal for the Philosophy of Science
Are there Quantum Jumps? (pp. 109–10)
Volume III, 1952

Shaw, George Bernard

Science becomes dangerous only when it imagines that it has reached its goal.

The Doctor's Dilemma
Preface
The Latest Theories (p. xc)

Shelley, Percy Bysshe

The cultivation of those sciences which have enlarged the limits of the empire of man over the external world, has, for want of the poetical faculty, proportionally circumscribed those of the internal world; and man, having enslaved the elements, remains himself a slave.

In Fanny Delisle
A Study of Shelley's A Defence of Poetry
Volume I, Line 1223 (p. 138)

Siegel, Eli

Science comes from the knowing that you want to know.

Damned Welcome
Aesthetic Realism Maxims
Part I, #298 (p. 68)

Smith, Adam

Science is the great antidote to the poison of enthusiasm and superstition.

The Wealth of Nations
Book V, Chapter I, Part III, Section III (p. 748)

Smith, Sydney

Science is his forte, and omniscience his foible.

In Isaac Todhunter
William Whewell
Volume I
Conclusion (p. 410)

Smithells, Arthur

Science has its roots and has gained its greatest impulse in the avocations of men.

In D.W. Hill
Science
Chapter 2 (p. 10)

Snow, C.P.

But after the idyllic years of science, we passed into a tempest of history; and by an unfortunate coincidence, we passed into a technological tempest, too.

In Paul C. Obler and Herman A. Estrin (Editors)
The New Scientist: Essays on the Methods and Value of Modern Science
The Moral Un-Neutrality of Science (p. 135)

Spark, Muriel

Art and religion first; then philosophy; lastly science. That is the order of the great subjects of life, that's their order of importance.

The Prime of Miss Jean Brodie
Chapter 2 (p. 39)

Spencer, Herbert

Only when Genius is married to Science, can the highest results be produced.

Education
Chapter I (p. 70)

Thus to the question with which we set out—What knowledge is of most worth?—the uniform reply is—Science. This is the verdict on all counts. For direct self-preservation, or the maintenance of life and health, the all-important knowledge is—Science. For the indirect self-preservation which we call gaining a livelihood, the knowledge of greatest value is—Science. For that interpretation of national life, past and present, without which the citizen cannot rightly regulate his conduct, the indispensable key is—Science. Alike for the most perfect production and highest enjoyment of art in all its forms, the needful preparation is still—Science. And for the purposes of discipline—intellectual, moral, religious—the most efficient study is, once more—Science.

Education
Chapter I (pp. 84–5)

Spencer-Brown, George

Left to itself, the world of science slowly diminishes as each result classed as scientific has to be reclassed as anecdotal or historical . . . Science is a continuous living process; it is made up of activities rather than records; and if the activities cease it dies.

Probability and Scientific Inference (p. 107)

Sterne, Laurence

Sciences may be learned by rote, but Wisdom not.

Tristram Shandy
Book V, Chapter 32 (p. 409)

Strutt, Robert John

There are some great men of science whose charm consists in having said the first word on a subject, in having introduced some new idea which has proved fruitful; there are others whose charm consists perhaps in having said the last word on the subject, and who have reduced the subject to logical consistency and clearness.

Life of John William Strutt: Third Baron Rayleigh
Chapter XVII (p. 310)

Sullivan, J.W.N.

Science, like everything else that man has created, exists, of course, to gratify certain human needs and desires. The fact that it has been steadily pursued for so many centuries, that it has attracted an ever-wider extent of attention, and that it is now the dominant intellectual interest of mankind, shows that it appeals to a very powerful and persistent group of appetites.

The Limitations of Science
Introduction (p. 7)

Swann, W.F.G.

Science walks hand in hand with human development as its constant benefactor, as the guardian of its peace, in a universe rich to provide happiness and security for all.

In D.W. Hill
Science
Chapter 3 (p. 37)

Temple, Frederick

The regularity of nature is the first postulate of Science; but it requires the very slightest observation to show us that, along with this regularity, there exists a vast irregularity which Science can only deal with by exclusion from its province.

The Relations between Religion and Science (p. 99)

Temple, G.

. . . any serious examination of the basic concepts of any science is far more difficult than the elaboration of their ultimate consequences.

Turning Points in Physics
From the Relative to the Absolute (p. 68)

Tennyson, Alfred Lord

Science moves, but slowly, slowly, creeping on from point to point.

The Complete Poetical Works of Tennyson
Locksley Hall
L. 134

. . . nourishing a youth sublime
With the fairy tales of science, and the long result of time.

The Complete Poetical Works of Tennyson
Locksley Hall
L. 11–12

Thompson, A.R.

Science has not only helped to destroy popular traditions that might have nourished a modern spirit of admiration, but has fostered a wintry skepticism, making man appear not an imperfect angel, but a super-educated monkey.

In R. Foerster (Editor)
Humanism and America
The Dilemma of Modern Tragedy (p. 129)

Thomson, J.A.

When science makes minor mysteries disappear, greater mysteries stand confessed. For one object of delight whose emotional value science has inevitably lessened—as Newton damaged the rainbow for Keats—science gives back double. To the grand primary impression of the world power, the immensities, the pervading order, and the universal flux, with which the man of feeling has been nutured from the old, modern science has added thrilling impressions of manifoldedness, intricacy, uniformity, inter-relatedness, and evolution. Science widens and clears the emotional window. There are great vistas to which science alone can lead, and they make for elevation of mind. The opposition between science and feeling is largely a misunderstanding. As one of our philosophers has remarked, Science is in a true sense "one of the humanities."

The Outline of Science
Volume IV
Science and Modern Thought (pp. 1176–7)

Science as science never asks the question *Why?* That is to say, it never inquires into the meaning, or significance, or purpose of this manifold Being, Becoming, and Having Been . . . Thus science does not pretend to be a bedrock of truth.

In Bertrand Russell
Religion and Science
Mysticism (p. 175)

Science is not wrapped up with any particular body of facts; it is characterized as an intellectual attitude. It is not tied down to any peculiar methods of inquiry; it is simply sincere critical thought, which admits conclusions only when these are based on evidence.

Introduction to Science
Chapter I (p. 27)

Thurber, James

Science has zipped the atom open in a dozen places, it can read the scrawling on the Rosetta stone as glibly as a literary critic explains Hart Crane, but it doesn't know anything about playwrights.

Collecting Himself
Roaming in the Gloaming (p. 194)

I Timothy 6:20

. . . keep that which is committed to thy trust, avoiding profane *and* vain babblings, and oppositions of science falsely so called . . .

The Bible

Tolstoy, Leo

What is called science to-day consists of a haphazard heap of information, united by nothing, often utterly unnecessary, and not only failing to present one unquestionable truth, but as often as not containing the grossest errors, today put forward as truths, and tomorrow overthrown.

What is Religion?
Chapter I (p. 3)

The highest wisdom has but one science—the science of the whole—the science explaining the whole creation and man's place in it.

War and Peace
Book V, Chapter 2 (p. 197)

Twain, Mark

But what we most admire is the vast capacity of that intellect which, without effort, takes in at once all the domains of science—all the past, the present and the future, all the errors of two thousand years, all the encouraging signs of the passing times, all the bright hopes of the coming age.

Is Shakespeare Dead?
Chapter X (p. 124)

Unamuno, Miguel de

Science exists only in personal consciousness and thanks to it; astronomy, mathematics, have no other reality than that which they possess as knowledge in the minds of those who study and cultivate them.

The Tragic Sense of Life
The Starting-Point (p. 31)

Science is a cemetery of dead ideas, even though life may issue from them.

The Tragic Sense of Life
The Rationalist Dissolutions (p. 90)

True science teaches, above all, to doubt and be ignorant.

The Tragic Sense of Life
The Rationalist Dissolutions (p. 93)

Science teaches us, in effect, to submit our reason to the truth and to know and judge of things as they are—that is to say, as they themselves choose to be and not as we would have them be.

The Tragic Sense of Life
Faith, Hope, and Charity (p. 197)

Wisdom is to science what death is to life, or, if you prefer it, wisdom is to death what science is to life.

Essays and Soliloquies
Some Arbitrary Reflections Upon Europeanization (p. 55)

Union Carbide and Carbon
More Jobs Through Science

Advertising Slogan

Unknown
The higher we soar on the wings of science,
the worse our feet seem to get entangled in the wires.

The New Yorker
February 7, 1931

Valéry, Paul
Science is feasible when the variables are few and can be enumerated; when their combinations are distinct and clear. We are tending toward the condition of science and aspiring to do it. The artist works out his own formulas; the interests of science lie in the *art* of making science.

The Collected Works of Paul Valéry
Volume 14
Analects (p. 191)

Vernadskii, V.I.
Science is alone and the routes to its achievement are alone. They are independent from the ideas of man, from his aspirations and wishes, from the social tenor of his life, from his philosophical, social, and religious theories. They are independent from his will and from his world outlook—they are primordial.

In Loren R. Graham
The Soviet Academy of Sciences and the Communist Party, 1927–1932
Chapter III (p. 80)

Virchow, Rudolf

. . . if we would serve science, we must extend her limits, not only as far as our own knowledge is concerned, but in the estimation of others.

Cellular Pathology
Author's Preface (p. 7)

There can be no scientific dispute with respect to faith, for science and faith exclude one another.

Disease, Life, and Man
On Man (p. 68)

The task of science, therefore, is not to attack the objects of faith, but to establish the limits beyond which knowledge cannot go and to found a unified self-consciousness within these limits.

Disease, Life, and Man
On Man (p. 69)

"Science in itself" is nothing, for it exists only in the human beings who are its bearers. "Science for its own sake" usually means nothing more than science for the sake of the people who happen to be pursuing it.

Disease, Life, and Man
Standpoints in Scientific Medicine (pp. 29–30)

Has not science the noble privilege of carrying on its controversies without personal quarrels?

Bulletin of the New York Academy of Medicine
Volume 4, 1928 (p. 995)

Voltaire

True science necessarily carries tolerance with it.

Correspondance de Voltaire
1881 edition
Volume XII
Letter to Madame d'Epinay
July 6, 1766 (p. 329)

Waddington, C.H.

. . . science, if given its head, is not just cold efficiency; its attitude is tolerant, friendly and humane. It has already become the dominant inspiration of human culture, so that modern poetry, painting and architecture derive their most constructive ideas from scientific thought.

The Scientific Attitude
The Scientific Attitude (p. 1)

Science is the organized attempt of mankind to discover how things work as causal systems.

The Scientific Attitude
Foreword (p. 9)

Science as a whole certainly cannot allow its judgment about facts to be distorted by ideas of what ought to be true, or what one may hope to be true.

The Scientific Attitude
Science Is Not Neutral (p. 25)

Watson, David Lindsay
Science sprawls over all the horizons of the modern mind like some vast cloudbank. The outlook and method of science penetrate relentlessly the strata of daily custom into the caverns of the unconscious mind itself. Science is by far the most powerful intellectual phenomenon of modern times, inexorably laying down the law in regions far from the laboratory, and subtly governing, by its techniques and devices, our modes of life and ways of thinking.

Scientists Are Human
Chapter I (p. 1)

The main vehicle of science is not the published formulations of laws and experiments in books and periodicals. This vehicle is, first and foremost, *men* who are worthy of them, who can understand and use the laws. But more than this: the vehicle is also the pattern of the society that can produce such men.

Scientists Are Human
Chapter I (p. 3)

Weaver, Warren
It is hardly necessary to argue, these days, that science is essential to the public. It is becoming equally true, as the support of science moves more and more to state and national sources, that the public is essential to science. The lack of general comprehension of science is thus dangerous both to science and the public, these being interlocked aspects of the common danger that scientists will not be given the freedom, the understanding, and the support that are necessary for vigorous and imaginative development.

In Hilary J. Deason
A Guide to Science Reading (p. 38)

Weber, Max
Science today is a 'vocation' organized in special disciplines in the service of self-clarification and knowledge of interrelated facts. It is not the gift of grace of seers and prophets dispensing sacred values and revelations,

nor does it partake of the contemplation of sages and philosophers about the meaning of the Universe.

In H.H. Gerth and C. Wright Mills (Editors)
From Max Weber
Science as a Vocation (p. 152)

Weil, Simone
To us, men of the West, a very strange thing happened at the turn of the century; without noticing it, we lost science, or at least the thing that had been called by that name for the last four centuries. What we now have in place of it is something different, radically different, and we don't know what it is. Nobody knows what it is.

On Science, Necessity, and the Love of God
Classical Science and After
Chapter I (p. 3)

Science today will either have to seek a source of inspiration higher than itself or perish. Science only offers three kinds of interest: 1. Technical applications. 2. A game of chess. 3. A road to God. (Attractions are added to the game of chess in the shape of competitions, prizes, and medals.)

Gravity and Grace
Intelligence (pp. 186–7)

Science is today regarded by some as a mere catalogue of technical recipes, and others as a body of pure intellectual speculations which are sufficient unto themselves; the former set too little value on the intellect, the latter on the world.

Oppression and Liberty
Theoretical Picture of a Free Society (pp. 104–5)

Weinberg, Steven
. . . there is an essential element in science that is cold, objective, and nonhuman . . . the laws of nature are as impersonal and free of human values as the rules of arithmetic . . . Nowhere do we see human value or human meaning.

Daedalus
Reflections of a Working Scientist
Volume 103, 1974 (p. 3)

Weiss, Paul A.
Science, to some, is Lady Bountiful, to others is the Villain of the Century. Some years ago, a book called it our "Sacred Cow," and certainly to many it has at least the glitter of the "Golden Calf." Glorification at one extreme, vituperation at the other.

Within the Gates of Science and Beyond
Science Looks at Itself (p. 25)

Weisskopf, Victor F.

Science developed only when men refrained from asking general questions such as: What is matter made of? How was the universe created? What is the essence of life? Instead they asked limited questions such as: How does an object fall? How does water flow in a tube? Thus, in place of asking general questions and receiving limited answers, they asked limited questions and found general answers. It remains a great miracle, that this process succeeded, and that the answerable questions became gradually more and more universal.

Science
The Significance of Science (p. 143)
Volume 176, Number 4031, April 14, 1972

Science is an important part of the humanities because it is based on an essential human trait: curiosity about how and why of our environment. We must foster wonder, joy of insight.

The Privilege of Being a Physicist
Chapter 4 (p. 33)

Whetham, W.C.D.

But beyond the bright search-light of science,
  Out of sight of the windows of sense,
Old riddles still bid us defiance,
  Old questions of Why and of Whence.
There fail all sure means of trial,
  There end all the pathways we've trod,
Where man, by belief or denial,
  Is weaving the purpose of God.

The Recent Development of Physical Science
Introduction (p. 10)

Whewell, William

Man is the interpreter of Nature, Science is the right interpretation.

The Philosophy of the Inductive Sciences
Volume II
Aphorisms Concerning Ideas
Aphorism I (p. 445)

Whitehead, Alfred North

Science is taking on a new aspect which is neither purely physical nor purely biological. It is becoming the study of organisms. Biology is the study of larger organisms; whereas physics is the study of the smaller organisms.

Science and the Modern World
Chapter VI (p. 103)

. . . science conceived as resting on mere sense-perception, with no other source of observation, is bankrupt, so far as concerns its claim to self-sufficiency. Science can find no individual enjoyment in nature: Science can find no aim in nature: Science can find no creativity in nature; it finds mere rules of succession. These negations are true of Natural Science. They are inherent in its methodology . . .

Modes of Thought
Chapter VIII (p. 211)

Science is even more changeable than theology.

Science and the Modern World
Chapter XII (p. 183)

Science is simply setting out on a fishing expedition to see whether it cannot find some procedure which it can call measurement of space and some procedure which it can call the measurement of time, and something which it can call a system of forces, and something which it can call masses, . . .

The Concept of Nature (p. 139–40)

The aim of science is to seek the simplest explanation of complex facts . . . Seek simplicity and distrust it.

The Concept of Nature (p. 163)

Science is *either* an important statement of systematic theory correlating observations of a common world *or* is the daydream of a solitary intelligence with a taste for the daydream of publication.

Process and Reality
Chapter V, Section IV (p. 502)

Science has always suffered from the vice of overstatement. In this way conclusions true within strict limitations have been generalized dogmatically into a fallacious universality.

The Function of Reason
Chapter I (p. 22)

There can be no true physical science which looks first to mathematics for the provision of a conceptual model. Such a procedure is to repeat the errors of the logicians of the middle ages.

Principles of Relativity (p. 39)

Aristotle discovered all the half-truths which were necessary to the creation of science.

Dialogues
Dialogue XLII

Science is the organisation of thought.

The Organisation of Thought
Chapter VI (p. 106)

A science which hesitates to forget its founders is lost.

The Organisation of Thought
Chapter VI (p. 115)

A man who only knows his own science, as a routine peculiar to that science, does not even know that. He has no fertility of thought, no power of quick seizing the bearing of alien ideas. He will discover nothing, and be stupid in practical applications.

The Organisation of Thought
Chapter II (p. 46)

Science is a river with two sources, the practical source and the theoretical source. The practical source is the desire to direct our actions to achieve predetermined ends . . . The theoretical source is the desire to understand.

The Organisation of Thought
Chapter VI (p. 106)

Science is in the minds of men but men sleep and forget, and at their best in any one moment of insight entertain but scanty thoughts. Science therefore is nothing but a confident expectation that relevant thoughts will occasionally occur.

The Principles of Natural Knowledge (p. 10)

Wilson, Edward O.
Important science is not just any similarity glimpsed for the first time. It offers analogues that map the gateways to unexplored terrain.

Biophilia
The Poetic Species (p. 67)

To a considerable degree science consists in originating the maximum amount of information with the minimum expenditure of energy. Beauty is the cleanness of line in such formulations along with symmetry, surprise, and congruence with other prevailing beliefs.

Biophilia
The Poetic Species (p. 60)

Winfree, Arthur
The basic idea of Western science is that you don't have to take into account the falling of a leaf on some planet in another galaxy when you're trying to account for the motion of a billiard ball on a pool table on earth. Very small influences can be neglected. There's a convergence

in the way things work, and arbitrarily small influences don't blow up to have arbitrarily large effects.

In J. Gleick
Chaos: Making a New Science (p. 15)

Wittgenstein, Ludwig
Man has to awaken to wonder—and so perhaps do peoples. Science is a way of sending him to sleep again.

Culture and Value (p. 5e)

Wolpert, Lewis
When we come to face the problems before us—poverty, pollution, overpopulation, illness—it is to science that we must turn, not to gurus. The arrogance of scientists is not nearly as dangerous as the arrogance that comes from ignorance.

New Scientist
In Mary Midgley
Can Science Save Its Soul? (p. 24)
August 1, 1992

Wordsworth, William
Science appears but what in truth she is,
Not as our glory and our absolute boast,
But as a succedaneum and a prop
To our infirmity. No officious slave
Art thou of that false secondary power
By which we multiply distinctions, then
Deem that our puny boundaries are things
That we perceive, and not that we have made.

The Complete Poetical Works of William Wordsworth
The Prelude
Book II, L. 212–219

Yates, Frances
Is not all science a gnosis, an insight into the nature of the All, which proceeds by successive revelations?

Giordano Bruno and the Hermetic Tradition (p. 452)

Ziman, John
In science, to echo Beethoven's dictum about music, 'Everything should be both surprising and expected'.

Reliable Knowledge
Chapter 3 (fn 17, p. 71)

Zinsser, Hans

Science is but a method. Whatever its material, an observation accurately made and free of compromise to bias and desire, and undeterred by consequence, is science.

The Atlantic Monthly
Untheological Reflections (p. 91)
July 1929

The life of a student of any science is a constant series of frustrations. From his own observations and those of others, a trellis of theory is built up beyond the solid stakes of fact. The investigator tests these, perched on scaffoldings of experiment which break down again and again and are, as often, reconstructed with the weak points reënforced. Eventually, as soon as he has tied down an elusive shoot, he loses interest and is lured by the ones a little higher up. There is never an end, and never a complete satisfaction—as there may be in the arts, when a perfect sonnet or a good statue is, in itself, final and forever.

As I Remember Him
Chapter XX (p. 330)

Should we force science down the throats of those that have no taste for it? Is it our duty to drag them kicking and screaming into the twenty-first century? I am afraid that it is.
Sir George Porter – (See p. 237)

SCIENCE AND ART

Bernstein, Jeremy

In science as in the arts, sound aesthetic judgments are usually arrived at only in retrospect. A really new art form or scientific idea is almost certain at first to appear ugly. The obviously beautiful, in both science and the arts, is more often than not an extension of the familiar. It is sometimes only with the passage of time that a really new idea begins to seem beautiful. As has been said, "Pioneers occupy new land. Only later, one comes to understand that the cabins they built were really cathedrals."

Experiencing Science
Chapter 1 (p. 3)

Blake, William

He who would do good to another must
do it in Minute Particulars:
General Good is the plea of the scoundrel,
hypocrite and flatterer;
For Art and Science cannot exist but in
minutely organized Particulars.

Jerusalem
The Holiness of Minute Particulars
3, Section 55 (p. 399)

Brown, John

Science and Art are the offspring of light and truth, of intelligence and will; they are the parents of philosophy—that its father, this its mother. Art comes up out of darkness, like a flower,—is there before you are aware, its roots unseen, not to be meddled with safely . . . It draws its nourishment from all its neighborhood, taking this and rejecting that, by virtue of its elective instinct knowing what is good for it; it lives upon the débris of former life. Science comes from the market; it is sold, can be measured and weighed, can be handled and gauged. It is full of light;

but is lucid rather than luminous; it is, at its best, food, not blood, much less muscle—the fuel, not the fire.

Horae Subsecivae
Series I
Art and Science

Campbell, Norman R.
Science, like art, should not be something extraneous, added as a decoration to other activities of existence; it should be part of them, inspiring our most trivial actions as well as our noblest thoughts.

What is Science?
Chapter VIII (p. 183)

Cassirer, Ernst
Since art and science move in entirely different planes, they cannot contradict or thwart one another.

An Essay on Man: An Introduction to a Philosophy of Human Culture
Chapter IX (p. 170)

Connolly, Cyril
Today the function of the artist is to bring imagination to science and science to imagination, where they meet, in the myth.

The Unquiet Grave
Part III (p. 114)

Delbrück, Max
The books of the great scientists are gathering dust on the shelves of learned libraries. And rightly so. The scientist addresses an infinitesimal audience of fellow composers. His message is not devoid of universality but its universality is disembodied and anonymous. While the artist's communication is linked forever with its original form, that of the scientist is modified, amplified, fused with the ideas and results of others, and melts into the stream of knowledge and ideas which forms our culture. The scientist has in common with the artist only this: that he can find no better retreat from the world than his work and also no stronger link with the world than his work.

Science
A Physicist's Renewed Look at Biology: Twenty Years Later (p. 1314)
Volume 168, Number 3937, 1970

Durant, William
Every science begins as philosophy and ends as art; it arises in hypothesis and flows into achievement.

The Story of Philosophy (p. 2)

Einstein, Albert

Science exists for Science's sake, like Art for Art's sake, and does not go in for special pleading or for the demonstration of absurdities.

Cosmic Religion
On Science (p. 100)

. . . one of the strongest motives that lead men to art and science is escape from everyday life with its painful crudity and hopeless dreariness, from the fetters of one's own ever shifting desires. A finely tempered nature longs to escape from personal life into the world of objective perception and thought . . . Man tries to make for himself in the fashion that suits him best a simplified and intelligible picture of the world; he then tries to some extent to substitute this cosmos of his for the world of experience, and thus to overcome it. This is what the painter, the poet, the speculative philosopher, and the natural scientist do, each in his own way.

The World As I See It
Principles of Research (pp. 20–1)

Escher, M.C.

. . . science and art sometimes can touch one another, like two pieces of the jigsaw puzzle which is our human life, and that contact may be made across the borderline between the two respective domains.

In Doris Schattschneider
Visions of Symmetry: Notebooks, Periodic Drawings, and Related Works of M.C. Escher
Chapter 2 (p. 104)

Harvey, William

On the same terms, therefore, as art is attained to, is all knowledge and science acquired; for as art is a habit with reference to things to be done, so is science a habit in respect to things to be known; as that proceeds from the initation of types or forms so this proceeds from the knowledge of natural things.

Anatomical Exercises on the Generation of Animals
Of the Manner and Order of Acquiring Knowledge (p. 333)

Heisenberg, Werner

Both science and art form in the course of the centuries a human language by which we can speak about the more remote parts of reality, and the coherent sets of concepts as well as the different styles of art are different words or groups of words in this language.

Physics and Philosophy
Chapter VI (p. 109)

Huxley, Aldous

Science and art are only too often a superior kind of dope, possessing this advantage over booze and morphia: that they can be indulged in with a good conscience and with the conviction that, in the process of indulging, one is leading the "higher life."

Ends and Means
Beliefs

Unlike art, science is genuinely progressive. Achievement in the fields of research and technology is cumulative; each generation begins at the point where its predecessor left off.

Science, Liberty and Peace (p. 30)

Karanikas, Alexander
. . . science pierces reality like a dagger in search of fact and truth while art caresses reality looking for pleasure, grace and beauty.

Tillers of a Myth
Science, The False Messiah (p. 127)

Kepes, Gyorgy
The essential vision of reality presents us not with fugitive appearances but with felt patterns of order which have coherence and meaning for the eye and for the mind. Symmetry, balance and rhythmic sequences express characteristics of natural phenomena: the connectedness of nature—the order, the logic, the living process. Here art and science meet on common ground.

The New Landscape in Art and Science
Chapter I (p. 24)

Klee, Paul
. . . the worst state of affairs is when science begins to concern itself with art.

The Diaries of Paul Klee 1898–1918
Diary III
Number 747 (p. 194)

Knuth, Donald
The difference between art and science is that science is what people understand well enough to explain to a computer. All else is art.

In Robert Slater
Portraits in Silicon
Chapter 31(p. 351)

Krauss, Karl
Science is spectrum analysis.
Art is photosynthesis.

In John D. Barrow
The Artful Universe (p. 114)

Santayana, George

Science is the response to the demand for information . . . Art is the response to the demand for entertainment.

The Sense of Beauty (p. 15)

Unknown

Art c'est moi—La Science c'est nous.
[Art is personal—Science is universal.]

In Pierre Lecomte du Noüy
The Road to Reason
Chapter 1 (p. 31)

Valéry, Paul

There is a *science* of simple things, an *art* of complicated ones. Science is feasible when the variables are few and can be enumerated; when their combinations are distinct and clear.

We are tending toward the condition of science and aspiring to it. The artist works out his own fornulas; the interest of science lies in the *art* of making science.

The Collected Works of Paul Valéry
Volume 14
Moralités
Analectes (p. 64)

Waddington, C.H.

The best of modern art is compatible only with true science, and the bogus science requires a fake art to keep it company.

The Scientific Attitude
Science Is Not Neutral (p. 27)

Whewell, William

Art and Science differ. The object of Science is Knowledge; the objects of Art, are Works. In Art, truth is a means to an end; in Science, it is the only end. Hence the Practical Arts are not to be classed among the Sciences.

The Philosophy of the Inductive Sciences
Volume II
Aphorisms Concerning Science
Aphorism XXV (p. 471)

Wordsworth, William

Enough of Science and of Art;
Close up these barren leaves;
Come forth, and bring with you a heart
That watches and receives.

The Complete Poetical Works of William Wordsworth
The Tables Turned
Stanza 8

SCIENCE AND FREEDOM

Eisenhower, Dwight D.

Science, great as it is, remains always the servant and the handmaiden of freedom. And a free science will ever be one of the most effective tools through which man will eventually bring to realization his age-old aspiration for an abundant life, with peace and justice for all.

In Dael Wolfle
Symposium on Basic Research
Science: Handmaiden of Freedom (p. 142)

Jefferson, Thomas

The main object of all science . . . was the freedom and happiness of man.

In Philip A. Bruce
History of the University of Virginia
Volume I
Introduction
Chapter IV (p. 31)

Oppenheimer, J. Robert

. . . as long as men are free to ask what they must, free to say what they think, free to think what they will, freedom can never be lost, and science can never regress.

In Lincoln Barnett
Life
J. Robert Oppenheimer (p. 133)
October 10, 1949

Sarnoff, David

Freedom is the oxygen without which science cannot breathe.

In Emily Davie (Editor)
Profile of America
Electronics—Today and Tomorrow (p. 232)

SCIENCE AND GOD

Bernal, J.D.
The role of God in the material world has been reduced stage by stage with the advance of science, so much so that He only survives in the vaguest mathematical form in the minds of older physicists and biologists.

In C.H. Waddington (Editor)
Science and Ethics
A Marxist Critique (p. 116)

Coulson, C.A.
. . . science is one aspect of God's presence, and scientists therefore part of the company of His heralds.

Science and Christian Belief
Scientific Method (p. 30)

du Noüy, Pierre Lecomte
Any man who believes in God must realize that no scientific fact, as long as it is true, can contradict God. Otherwise, it would not be true. Therefore, any man who is afraid of science does not possess a strong faith.

Human Destiny
Chapter 16 (p. 243)

Fosdick, Harry Emerson
What modern science is doing for multitudes of people, as anybody who watches American life can see, is not to disprove God's theoretical existence, but to make him "progressively less essential."

Adventurous Religion
Will Science Displace God? (p. 136)

Garman, Charles E.
Science is thinking God's thoughts after Him just as truly as when we read the scriptures.

Letters, Lectures, Addresses
Science and Theism (p. 231)

Pope Pius XII
The more true science advances, the more it discovers God, almost, as though he were standing, vigilant behind every door which science opens.

Address
November 22, 1951

Stace, W.T.
. . . no scientific argument—by which I mean an argument drawn from the phenomena of nature—can ever have the slightest tendency either to prove or disprove the existence of God . . . science is irrelevant to religion.

Religion and the Modern Mind
The Consequences for Religion (p. 76)

Concepts are games we play with our heads;
methods are games we play with our hands . . .
Frank E. Egler – (See p. 30)

SCIENCE AND MORALS

Bronowski, Jacob
It is not the business of science to inherent the earth, but to inherit the moral imagination; because without that man and beliefs and science will perish together.

The Ascent of Man
Chapter 13

Chesterton, G.K.
. . . when any part of the general public is drawn into a debate on physical science, we may be certain that it has already become a debate on moral science.

All is Grist
On Gossip about Heredity (p. 96)

Compton, Karl Taylor
I would emphasize the fact that scientific discovery is, *per se*, neither good nor bad. It simply produces knowledge and with knowledge, opportunity and responsibility. I think it fair to say that the advance of science carries with it powerful demands on morality if the results are to be beneficial rather than harmful.

A Scientist Speaks (p. 5)

Dewey, John
Science through its physical technological consequences is now determining the relations which human beings, severally and in groups, sustain to one another. If it is incapable of developing moral techniques which will also determine these relations, the split in modern culture goes so deep that not only democracy but all civilized values are doomed.

Freedom and Culture
Chapter Six (p. 154)

Diderot, Denis

The moral universe is so closely linked to the physical universe that it is scarcely likely that they are not one and the same machine.

Éléments de Physiologie (pp. xiii–xiv)

Ferré, Nels F.S.

Science can be and is being made into an escapist philosophy—into a dodge of moral disciplines and spiritual responsibilities.

Faith and Reason
Chapter II (p. 83)

Friedenberg, Edgar Z.

. . . only science can hope to keep technology in some sort of moral order.

The Vanishing Adolescent
The Impact of the School
The Clarification of Experience (p. 50)

Huxley, Thomas

We men of science, at any rate, hold ourselves morally bound to "try all things and hold fast to that which is good"; and among public benefactors we reckon him who explodes old error, as next in rank to him who discovers new truth.

In F. Darwin
The Life and Letters of Charles Darwin
Volume III (p. 18)

Jefferson, Thomas

If science produces no better fruits than tyranny, murder, rapine and destitution of national morality, I would rather wish our country to be ignorant, honest and estimable, as our neighboring savages are.

Letter to John Adams
1812

Kruyt, H.R.

Clearer than ever we understand that knowledge is not all, that we need morals and brotherhood to avoid science becoming a curse.

International Council of Scientific Unions
First General Assembly following the Second World War
In John P. Dickinson
Science and Scientific Researchers in Modern Society (p. 165)

Lerner, Max

Science itself is a humanist in the sense that it doesn't discriminate between human beings, but it is also morally neutral. It is no better or worse than the ethos with and for which it is used.

New York Post
Manipulating Life
January 24, 1968

Masters, William H.

Science by itself has no moral dimension. But it does seek to establish truth. And upon this truth morality can be built.

Life
Two Sex Researchers on the Firing Line (p. 49)
June 24, 1966

Poincaré, Henri

There can no more be immoral science than there can be scientific morals.

The Foundations of Science
The Values of Science
Introduction (p. 206)

Snow, C.P.

. . . there is a moral component right in the grain of science itself . . .

The Two Cultures: and A Second Look
Chapter I (p. 19)

Toynbee, Arnold J.

Our western science is a child of moral virtues; and it must now become the father of further moral virtues if its extraordinary material triumphs in our time are not to bring human history to an abrupt, unpleasant and discreditable end.

The New York Times Magazine
A Turning Point in Man's Destiny (p. 5)
December 26, 1954

Wallace, Henry A.

I can understand the impulse which prompts scientists to defend science against the attacks of the uninformed. Science has achieved so many miracles for society, saved so many lives, made possible so extraordinary an advance in material living standards for so many millions of people, that it is disquieting to think that all the consequences of science can ever be other than good. Yet I don't see what basis we have for assuming that science can and does have only beneficial consequences. Is the product of man's curiosity inevitably good? Is there any reason for assuming that the end result of any enlargement of human knowledge must, perforce, be beneficial? It may be disturbing to realize it, but the truth seems to be that science proceeds without moral obligations; it is neither moral nor immoral, but in essence amoral.

Scientific Monthly
Scientists in an Unscientific Society (p. 285)
Volume 150, 1934

SCIENCE AND PHILOSOPHY

Chesterton, G.K.

To mix science up with philosophy is only to produce a philosophy that has lost all its ideal value and a science that has lost all its practical value.

All Things Considered
Science and Religion (p. 187)

Durant, William

Philosophy . . . is the front trench in the siege of truth. Science is the captured territory.

The Story of Philosophy
Introduction (p. 2)

Science is analytical description, philosophy is synthetic interpretation. Science wishes to resolve the whole into the known.

The Story of Philosophy
Introduction (p. 3)

Science without philosophy, facts without perspective and valuation, cannot save us from havoc and despair. Science gives us knowledge, but only philosophy can give us wisdom.

The Story of Philosophy
Introduction (p. 3)

Fabing, Harold
Marr, Ray

Not fact-finding, but attainment to philosophy, is the aim of science.

Fischerisms (p. 7)

Gornick, Vivian

Science—like art, religion, political theory, or psychoanalysis—is work that holds out the promise of philosophic understanding, excites in us the belief that we can "make sense of it all."

Women in Science
Part One (p. 66)

Jeans, Sir James

The philosophy of any period is always largely interwoven with the science of the period, so that any fundamental change in science must produce reactions in philosophy.

Physics and Philosophy
Chapter I (p. 2)

In whatever ways we define science and philosophy their territories are contiguous; wherever science leaves off—and in many places its boundary is ill-defined—there philosophy begins.

Physics and Philosophy
Chapter I (p. 17)

Science came to recognize that its only proper objects of study were the sensations that the objects of the external universe produced on our senses. The dictum *esse est percipi* was adopted whole-heartedly from philosophy—not because scientists had any predilection for an idealist philosophy, but because the assumption that things existed which could not be perceived had led them into a whole morass of inconsistencies and impossibilities. Those who did not adopt it were simply left behind, and the torch of knowledge was carried onwards by those who did.

Philosophy
The Mathematical Aspect of the Universe (p. 11)
Volume VII, Number 25, January 1932

Jones, Steve

. . . philosophy is to science as pornography is to sex.

New Scientist
In Mary Midgley
Can Science Save Its Soul? (p. 25)
August 1, 1992

Keats, John

Do not all charms fly
At the mere touch of cold philosophy?
There was an awful rainbow once in heaven:
We know her woof, her texture; she is given
In the dull catalogue of common things.
Philosophy will clip an Angel's wings,
Conquer all mysteries by rule and line,
Empty the haunted air, and gnom'ed mine—
Unweave a rainbow . . .

Complete Poems
Lamia, Part II, L. 229–237

Mercier, André

Philosophy does not 'solve problems', whereas science does. Philosophy, in its relations to science, gathers up the problems of science, which are no longer problems since they have found solutions, and seeks to order them in such a way that the structure of knowledge does, in fact, appear.

Nature
Fifty Years of the Theory of Relativity (p. 919)
Volume 175, May 28, 1955

Pope Pius XII

Science descends ever more deeply into the hidden recesses of things, but it must halt at a certain point when questions arise which cannot be settled by means of sense observations. At that point the scientist needs a light which is capable of revealing to him truth which entirely escapes his senses. This light is philosophy.

In Philip G. Fothergill
Life and Its Origin (p. 12)
Pontifical Academy of Science
Meeting 1955

Renan, Ernest

Socrates founded philosophy, and Aristotle science. There was philosophy before Socrates, and science before Aristotle; and since Socrates and since Aristotle, philosophy and science have made immense progress: but all has all been built upon the foundations they laid.

Life of Jesus
Chapter 28 (p. 411)

Russell, Bertrand

The man who has no tincture of philosophy goes through life imprisoned in the prejudices . . . To such a man the world tends to become definite, finite, obvious; common objects rouse no questions, and unfamiliar possibilities are contemptuously rejected.

The Problems of Philosophy
Chapter XV (pp. 156–7)

Unknown

. . . the philosophy of science is just about as useful to scientists as ornithology is to birds.

In S. Weinberg
Nature
Newtonianism, Reductionism and the Art of
Congressional Testimony (p. 433)
Volume 330, December 3, 1987

Whitehead, Alfred North

Philosophy is the product of wonder.

Nature and Life
Chapter I (p. 1)

The primary task of a philosophy of natural science is to elucidate the concept of nature, considered as one complex fact for knowledge, to exhibit the fundamental entities and the fundamental relations between entities in terms of which all laws of nature have to be stated, and to secure that the entities and relations thus exhibited are adequate for the expression of all the relations between entities which occur in nature.

The Concept of Nature
Chapter II (p. 46)

. . . [the scientist] is greeted by a band of theologians who have been sitting there for centuries.
Robert Jastrow – (See p. 284)

SCIENCE AND POETRY

Beston, Henry
Poetry is as necessary to comprehension as science. It is as impossible to live without reverence as it is without joy.

The Outermost House

Davis, Joel
Poetry and science are closer than most people realize. Many poets and scientists already know this, of course. Most of the rest of us are still trapped in dismal stereotypes about both fields of human endeavor. The deep link between the two is *vision*.

Alternate Realities
In a Grain of Sand (p. 3)

Day Lewis, C.
Science is concerned with finding out and stating the facts: poetry's task is to give you the look, the smell, the taste, the 'feel' of those facts.

Poetry for You
Chapter I (p. 10)

Every good poem, in fact, is a bridge built from the known, familiar side of life over into the unknown. Science, too, is always making expeditions into the unknown. But this does not mean that science can supersede poetry. For poetry enlightens us in a different way from science: it speaks directly to our feelings or imagination. The findings of poetry are no more and no less true than science.

Poetry for You
Chapter VIII (p. 92)

Jones, Frederick Wood
Whoever wins to a great scientific truth will find a poet before him in the quest.

Medical Journal of Australia
August 29, 1931

Schlegel, Friedrich von

Strictly understood, the concept of a scientific poem is quite as absurd as that of a poetical science.

Philosophical Fragments
Critical Fragments, 61 (p. 8)

Science is advanced by husbands, but wives are often behind them.
Alexis Romanoff – (See p. 241)

SCIENCE AND POLITICS

Budworth, D.

Science policy is essentially about the allocation of scarce resources, and is therefore a part of politics . . . The scarce resource with which science policy should concern itself in the short term is not money, but that portion of the scientific population which is capable of initiating and leading significant work. Such people are always in short supply, even when the total population itself is greater than the available jobs.

New Scientist
Science Policy Should Be About People (pp. 684–5)
Volume 69, Number 993, March 25, 1976

Johnson, Harry G.

Basic science, per se, contributes to culture; it contributes to our social well-being, including national defence and public health; to our economic well-being; and it is an essential element of the education not only of scientists but also of the population as a whole. In deciding how much science the society needs, one must decide how the support of science bears on these other, politically defined, goals of the society.

In National Academy of Sciences
Basic Research and National Goals:
A Report to the Committee on Science and Astronautics
Federal Support of Basic Research: Some Economic Issues
Summary (p. 5)

Koestler, Arthur

No scientist is admired for failing in the attempt to solve problems that lie beyond his competence. The most he can hope for is the kindly contempt earned by the Utopian politician. If politics is the art of the possible, research is surely the art of the soluble. Both are immensely practical-minded affairs.

New Statesman
The Act of Creation
Volume 19, June 1964

Maupertuis, Pierre Louis Moreau de

There are sciences over which the will of kings has no immediate influence; it can procure advancement there only in so far as the advantages which it attaches to their study can multiply the number and the efforts of those who apply themselves to them. But there are other sciences which for their progress urgently need the power of sovereigns; they are all those which require greater expenditure than individuals can make or experiments which would not ordinarily be practicable.

Lettres sur le progrès des sciences, Oeuvres de Maupertuis
Dresden (pp. 6–7)

Price, Don K.

. . . it has begun to seem evident to a great many administrators and politicians that science had become something very close to an *establishment*, in the old and proper sense of that word: a set of institutions supported by tax funds but largely on faith and without direct responsibility to political control.

The Scientific Estate
Chapter 1 (p. 12)

. . . all sciences are considered by their professors, as equally significant; by the politicians, as equally incomprehensible; and by the military as equally expensive.

The Scientific Estate
Chapter 1 (p. 12)

Rabinowitch, Eugene

Science has assumed such an important role in determining the parameters of national and international life, that participation in national decisions by people whose world picture has been affected by the study and practice of science (even if this picture has its own bias), is indispensable for many major political decisions—to correct the bias of the more traditional molders of national decisions, such as men with legal training.

The New Republic
Open Season on Scientists (p. 21)
January 1, 1966

Shakespeare, William

I have done the state some service, and they know it.

Othello, the Moor of Venice
Act V, Scene ii, L. 338

SCIENCE AND RELIGION

Berger, Peter L.

Protestant theologians have been increasingly engaged in playing a game whose rules have been dictated by their cognitive antagonists.

A Rumor of Angels
Chapter 1 (p. 12)

Buck, Pearl S.

Science and religion, religion and science, put it as I may they are two sides of the same glass, through which we see darkly until these two, focusing together, reveal the truth.

A Bridge for Passing
III (p. 255)

Bultmann, R.

. . . the New Testament provides a world picture which belongs entirely to Jewish or Gnostic mythology and is incredible or even meaningless in a scientific age.

In H.J. Paton
The Modern Predicament
Chapter XV, Section 3 (p. 228)

Burroughs, John

The mysteries of religion are of a different order from those of science; they are parts of an arbitrary system of man's own creation; they contradict our reason and our experience, while the mysteries of science are revealed by our reason, and transcend our experience.

The Atlantic Monthly
Scientific Faith (p. 33)
July 1915

The miracles of religion are to be discredited, not because we cannot conceive of them, but because they run counter to all the rest of our knowledge; while the mysteries of science, such as chemical affinity, the conservation of energy, the indivisibility of the atom, the change of the non-living into the living . . . extend the boundaries of our knowledge, though the *modus operandi* of the changes remains hidden.

The Atlantic Monthly
Scientific Faith (p. 33)
July 1915

Bush, Vannevar
To pursue science is not to disparage the things of the spirit. In fact, to pursue science rightly is to furnish a framework on which the spirit may rise.

Speech
Massachusetts Institute of Technology
October 5, 1953

Bushnell, Horace
As the science of nature goes toward completion, religion, having all the while been watching for it in close company, will have gotten immense breadth and solidity, from the ideas and facts unfolded in its discoveries, and will be as much enlarged in its confidence and the sentiment of worship, as beholding God's deep system in the world signifies more than looking on its surfaces.

Putnam's Magazine
Science and Religion (p. 267)
Volume 1, 1868

What is science, anyhow, but the knowledge of species? And if species do not keep their places, but go a masking or really becoming one another, in strange transmutations, what is there to know, and where is the possibility of science? If there is no stability or fixity in species, then, for aught that appears, even science itself may be transmuted into successions of music, and moonshine, and auroral fires. If a single kind is all kinds, then all are one, and since that is the same as none, there is knowledge no longer. The theory may be true, but it never can be proved, for that reason if no other. And when it is proved, if that must be the fact, we may well enough agree to live without religion.

Putnam's Magazine
Science and Religion (p. 271)
Volume 1, 1868

Clark, W.C.
Majone, G.

The social uses of science have always had something in common with the social uses of religion. And in the two decades following the Second World War, modern science took on a most religious-looking numinous legitimacy as an unquestioned source of authority on all manner of policy problems.

Report of the International Institute of Applied Systems Analysis
The Critical Appraisal of Scientific Inquiries with Policy Implications
Laxenburg, Austria, 1984 (p. 35)

Compton, Karl Taylor

Science has contributed to the making of religion into a developing dynamic spiritual force.

A Scientist Speaks (p. 19)

I Corinthians 15:46

That was not first which is spiritual, but that which is natural; and afterward that which is spiritual.

The Bible

Dobzhansky, Theodosius

There are still many people who are happy and comfortable adhering to fundamentalist creeds. This should cause no surprise, since a large majority of these believers are as unfamiliar with scientific findings as were people who lived centuries ago.

The Biology of Ultimate Concern
Chapter 5 (p. 95)

. . . nothing gives more pleasure to a rather common type of religious person than to point out that science cannot explain this or cannot account for that!

The Biology of Ultimate Concern
Chapter 5 (p. 97)

Science and religion deal with different aspects of existence . . . [T]hese are the aspect of facts and the aspect of meaning. But there is one stupendous fact . . . the meaning of which they have ceaselessly tried to discover. This fact is Man.

The Biology of Ultimate Concern
Chapter 5 p. 96)

Dyson, Freeman

Professional scientists today live under a taboo against mixing science and religion.

Disturbing the Universe
Chapter 23 (p. 245)

Eddington, Sir Arthur Stanley

The starting-point of belief in mystical religion is a conviction of significance or, as I have called it earlier, the sanction of a striving in the consciousness. This must be emphasised because appeal to intuitive conviction of this kind has been the foundation of religion through all ages, and I do not wish to give the impression that we have now found something new and more scientific to substitute. I repudiate the idea of proving the distinctive beliefs of religion either from the data of physical science or by the methods of physical science.

The Nature of the Physical World
Chapter XV (p. 333)

It is probably true that the recent changes of scientific thought remove some of the obstacles to a reconciliation of religion with science; but this must be carefully distinguished from any proposal to base religion on scientific discovery. For my own part I am wholly opposed to any such attempt.

Science and the Unseen World
Section VII (pp. 72–3)

Einstein, Albert

Science without religion is lame, religion without science is blind.

Out of My Later Years
Science and Religion
II (p. 26)

. . . science not only purifies the religious impulse of the dross of its anthropomorphism but also contributes to a religious spiritualization of our understanding of life.

Out of My Later Years
Science and Religion
II (p. 29)

The basis of all scientific work is the conviction that the world is an ordered and comprehensive entity, which is a religious sentiment. My religious feeling is a humble amazement at the order revealed in the small patch of reality to which our feeble intelligence is equal.

Cosmic Religion
On Science (p. 98)

By furthering logical thought and a logical attitude, science can diminish the amount of superstition in the world. There is no doubt that all but the crudest scientific work is based on a firm belief—akin to religious feeling—in the rationality and comprehensibility of the world.

Cosmic Religion
On Science (p. 98)

. . . the cosmic religious experience is the strongest and noblest driving force behind scientific research.

Cosmic Religion
Cosmic Religion (p. 52)

I am of the opinion that all the finer speculations in the realm of science spring from a deep religious feeling, and that without such feeling they would not be fruitful.

Forum
Science and God: A Dialog (p. 373)
Volume 83, June 1930

I have never found a better expression than "religious" for this trust in the rational nature of reality and of its peculiar accessibility to the human mind. Where this trust is lacking science degenerates into an uninspired procedure. Let the devil care if the priests make capital out of this. There is no remedy for that.

Lettres à Maurice Solovine (pp. 102–3)

Certain it is that a conviction, akin to religious feeling, of the rationality or intelligibility of the world lies behind all scientific work of a higher order . . . This firm belief, a belief bound up with deep feeling, in a superior mind that reveals itself in the world of experience, represents my conception of God.

Ideas and Opinions (p. 255)

Scientific research is based on the idea that everything that takes place is determined by laws of nature, and therefore this holds for the actions of people. For this reason, a research scientist will hardly be inclined to believe that events could be influenced by a prayer, i.e. by a wish addressed to a supernatural Being.

However, it must be admitted that our actual knowledge of these laws is only imperfect and fragmentary, so that, actually, the belief in the existence of basic all-embracing laws in Nature also rests on a sort of faith. All the same this faith has been largely justified so far by the success of scientific research.

But, on the other hand, every one who is seriously involved in the pursuit of science becomes convinced that a spirit is manifest in the laws of the Universe—a spirit vastly superior to that of man, and one in the face of which we with our modest powers must feel humble. In this way the pursuit of science leads to a religious feeling of a special sort, which is indeed quite different from the religiosity of someone more naive.

Letter dated January 24, 1936
Quoted in Helen Dukas and Banesh Hoffmann
Albert Einstein: The Human Side (pp. 32–4)

Emerson, Ralph Waldo

The Religion that is afraid of science dishonors God & commits suicide.

The Journals and Miscellaneous Notebooks of Ralph Waldo Emerson
Volume III
1826–1832
March 4, 1831 (p. 239)

Flaubert, Gustave

Science. A little science takes your religion from you; a great deal brings you back to it.

Dictionary of Accepted Ideas

Froude, James Anthony

The superstitions of science scoff at the superstitions of faith.

Eclectic Review
February 1852
The Lives of the Saints

Gilkey, Langdon

It is because science is limited to a certain level of explanation that scientific and religious theories can exist side by side without excluding one another, that one person can hold both to the scientific accounts of origins and to a religious account, to the creation of all things by God . . .

Creationism on Trial: Evolution and God at Little Rock
Chapter 5 (p. 117)

Goodspeed, Edgar J.

Science sees meaning in every part; religion sees meaning in the whole.

The Four Pillars of Democracy
Chapter V (p. 106)

. . . religion needs science, to protect it from religion's greatest danger, superstition.

The Four Pillars of Democracy
Chapter V (p. 115)

Science needs religion, to prevent it from becoming a curse to mankind instead of a blessing.

The Four Pillars of Democracy
Chapter VI (p. 134)

Grinnel, Frederick

. . . modern science constitutes a method for understanding and modifying the world but has no inherent direction, whereas modern religion describes a messianic world view but lacks a useful method to bring about this state of affairs.

Perspectives in Biology and Medicine
Complementarity: An Approach to Understanding the Relationship between Science and Religion (p. 293)
Volume 29, Number 2, Winter 1986

Gull, Sir William Withey

Realize, if you can, what a paralysing influence on all scientific inquiry the ancient belief must have had which attributed the operations of nature to the caprice not of one divinity only, but of many. There still remains vestiges of this in most of our minds, and the more distinct in proportion to our weakness and ignorance.

British Medical Journal
Volume 2, 1874 (p. 425)

Heisenberg, Werner

In the history of science, ever since the famous trial of Galileo, it has repeatedly been claimed that scientific truth cannot be reconciled with the religious interpretation of the world. Although I am now convinced that scientific truth is unassailable in its own field, I have never found it possible to dismiss the content of religious thinking as simply part of an outmoded phase in the consciousness of mankind, a part we shall have to give up from now on. Thus in the course of my life I have repeatedly been compelled to ponder on the relationship of these two regions of thought, for I have never been able to doubt the reality of that to which they point.

Across the Frontiers
Chapter XVI (p. 213)

Hertz, Rabbi Richard

I find no conflict between science and religion. Science teaches what is. Religion teaches what ought to be. Science describes. Religion prescribes. Science analyzes what we can see. Religion deals with what is unseen. Each can help the other.

The American Jew in Search of Himself
Chapter 4 (p. 42)

Hooykaas, R.

Metaphorically speaking, whereas the bodily ingredients of science may have been Greek, its vitamins and hormones were biblical.

Religion and the Rise of Modern Science (p. 162)

Huxley, Julian

. . . it is no longer possible to maintain that science and religion must operate in thought-tight compartments or concern separate sectors of life; they are both relevant to the whole of human existence.

In Teilhard de Chardin
The Phenomenon of Man
Introduction (p. 26)

Like the meridians as they approach the poles, science, philosophy and religion are bound to converge as they draw nearer to the whole. I say "converge" advisedly, but without merging, and without ceasing, to the very end, to assail the real from different angles and on different planes.

In Teilhard de Chardin
The Phenomenon of Man
Introduction (p. 30)

Huxley, Thomas

True science and true religion . . . are twin-sisters, and the separation of either from the other is sure to prove the death of both. Science prospers exactly in proportion as it is religious; and religion flourishes in exact proportion to the scientific depth and firmness of its basis. The great deeds of philosophers have been less the fruit of their intellect than of the direction of that intellect by an imminently religious tone of mind. Truth has yielded herself rather to their patience, their love, their single-heartedness, and their self-denial, than to their logical acumen.

In Herbert Spencer
Education
Chapter I (p. 81)

Inge, William Ralph

No scientific discovery is without its religious and moral implications.

Outspoken Essays
Confessio Fidei (p. 56)

Jastrow, Robert

For the scientist who has lived by his faith in the power of reason, the story ends like a bad dream. He has scaled the mountains of ignorance; he is about to conquer the highest peak; as he pulls himself over the final rock, he is greeted by a band of theologians who have been sitting there for centuries.

God and the Astronomers
The Religion of Science (pp. 105–6)

Joad, C.E.M.

. . . the philosophising of the physicists is noticeably inferior to their physics, and eminent men are at the moment engaged in making all the mistakes which the philosophers made for themselves some three hundred years ago and have been engaged in detecting and correcting ever since. In particular it is thought that modern physics lends support to Idealism, and suggests, if it does not actually require, a religious interpretation of the universe.

Guide to Modern Thought
Chapter I (p. 15–6)

Kaempffert, Waldemar

Religion may preach the brotherhood of man; science practices it.

In Edward R. Murrow
This I Believe: 2
Michael Faraday (p. 196)

King, Martin Luther, Jr.

Science investigates; religion interprets. Science gives man knowledge which is power; religion gives man wisdom which is control.

Strength to Love
Chapter I (p. 3)

McKenzie, John L.

Happily, we have survived into a day when science and theology no longer speak to each other in the language of fishmongers.

The Two-Edged Sword
Cosmic Origins (p. 74)

Mencken, H.L.

The effort to reconcile science and religion is almost always made, not by theologians, but by scientists unable to shake off altogether the piety absorbed with their mother's milk.

Minority Report
Number 232 (p. 166)

To me the scientific point of view is completely satisfying, and it has been so as long as I can remember. Not once in this life have I ever been inclined to seek a rock and a refuge elsewhere. It leaves a good many dark spots in the universe, to be sure, but not a hundredth time as many as theology. We may trust it, soon or late, to throw light upon many of them, and those that remain dark will be beyond illumination by any other agency. It also fails on occasion to console, but so does theology . . .

In Charles A. Fecher
Mencken: A Study of His Thought (p. 84)

Moore, John A.

A fundamental difference between religious and scientific thought is that the received beliefs in religion are ultimately based on revelations or pronouncements, usually by some long dead prophet or priest . . . Dogma is interpreted by a caste of priests and is accepted by the multitude on faith or under duress. In contrast, the statements of science are derived from the data of observations and experiment, and from the manipulation of these data according to logical and often mathematical procedures.

Science as a Way of Knowing: the Foundations of Modern Biology
Chapter 4 (p. 59)

Paley, William

There cannot be design without a designer; contrivance without a contriver; order without choice; arrangement, without any thing capable of arranging; subserviency and relation to a purpose, without that which could intend a purpose; means suitable to an end, without the end ever having been contemplated, or the means accommodated to it. Arrangement, disposition of parts, subserviency of means to an end, relation of instruments to an use, imply the presence of intelligence and mind.

Natural Theology
Chapter II (pp. 15–6)

Planck, Max

Religion belongs to that realm that is inviolable before the law of causation and therefore closed to science.

Where is Science Going?
The Answer of Science (p. 168)

There can never be any real opposition between religion and science; for the one is the compliment of the other.

Where is Science Going?
The Answer of Science (p. 168)

Religion and natural science . . . are in agreement, first of all, on the point that there exists a rational world order independent from man, and secondly, on the view that the character of this world order can never be directly known but can only be indirectly recognized or suspected. Religion employs in this connection its own characteristic symbols, while natural science uses measurements founded on sense experiences.

Scientific Autobiography and Other Papers
Religion and Natural Science
Part IV (pp. 182–3)

Religion and natural science are fighting a joint battle in an incessant, never relaxing crusade against skepticism and against dogmatism, against disbelief and against superstition, and the rallying cry in this crusade has always been, and always will be, "*On to God*."

Scientific Autobiography and Other Papers
Religion and Natural Science
Part IV (p. 187)

Polanyi, M.

Admittedly, religious conversion commits our whole person and changes our whole being in a way that an expansion of natural knowledge does not do. But once the dynamics of knowing are recognized as the dominant principle of knowledge, the difference appears only as one of degree . . . it established a continuous ascent from our less personal knowing of inanimate matter to our convivial knowing of living beings and beyond this to knowing our responsible fellow men. Such I believe is the true transition from the sciences to the humanities and also from our knowing the laws of nature to our knowing the person of God.

Journal of Religion
Faith and Reason (pp. 244, 245)
Volume XLI, Number 4, October 1961

Pope John Paul II

Science can purify religion from error and superstition. Religion can purify science from idolatry and false absolutes.

In James Reston
Galileo, A Life (p. 461)

Popper, Karl R.

Science is most significant as one of the greatest spiritual adventures that man has yet known.

In John Oulton Wisdom
Foundations of Inference in Natural Science (p. v)

Raven, Charles E.

To mention Science and Religion in the same sentence is . . . to affirm an antithesis and suggest a conflict.

Science, Religion and the Future
Chapter I (p. 1)

Reichenbach, Hans

The belief in science has replaced in large measure, the belief in God. Even where religion was regarded as compatible with science, it was modified by the mentality of the believer in scientific truth.

The Rise of Scientific Philosophy
Chapter 3 (p. 44)

Roelofs, Howard Dykema

Religion can produce on occasion what science never does, namely, saints. Today we have science and scientists aplenty. We lack saints.

In Herbert J. Muller
Science and Criticism
Chapter III (p. 59)

Sa'di

Science is for the cultivation of religion, not for worldly enjoyment.

The Gulistān
Chapter VIII
Maxim IV (p. 206)

Shaw, George Bernard

Let the Churches ask themselves why there is no revolt against the dogmas of mathematics though there is one against the dogmas of religions. It is not that the mathematical dogmas are more comprehensible . . . It is not that science is free from legends, witchcraft, miracles, biographic boostings of quacks as heroes and saints, and of barren scoundrels as explorers and discoverers . . . But no student of science has yet been taught that specific gravity consists in the belief that Archimedes jumped out of the bath and ran naked through the streets of Syracuse shouting Eureka, Eureka, or that the law of inverse squares must be discarded if anyone can prove that Newton was never in an orchard in his life.

Back to Methuselah
Preface
lxxxix

Sperry, Roger

Probably the widest, deepest rift in contemporary culture and the source of its most profound conflict is that separating the two major opposing views of existence upheld by science and by orthodox religions, respectively. Together they represent two totally different kinds of 'truth', the former asking us to accept impersonal mass–energy accounts of the cosmos, the latter requiring faith in varied spiritual explanations.

Perspectives in Biology and Medicine
The New Mentalist Paradigm and Ultimate Concern (p. 415)
Volume 29, Number 3, Part I, Spring 1986

Streeter, B.H.

Science is the great cleanser of the human thinking; it makes impossible any religion but the highest.

Reality
Chapter IX (p. 272)

Teilhard de Chardin, Pierre
Religion and science are the two conjugated faces of phases of one and the same act of complete knowledge—the only one which can embrace the past and future of evolution so as to contemplate, measure and fulfill them.

The Phenomenon of Man
Chapter Three, Section 2 (p. 285)

Temple, Frederick
Science and Religion seem very often to be the most determined foes to each other that can be found. The scientific man often asserts that he cannot find God in Science; and the religious man often asserts that he cannot find Science in God.

The Relations between Religion and Science (p. 4)

Science postulates uniformity; Religion postulates liberty.

The Relations between Religion and Science (p. 70)

Tillich, Paul
. . . theology cannot rest on scientific theory. But it must relate its understanding of man to an understanding of universal nature, for man is a part of nature and statements about nature underlie every statement about him.

In T. Dobzhansky
The Biology of Ultimate Concern
Chapter 6 (pp. 109–10)

Toynbee, Arnold J.
Theology, not religion, is the antithesis to science.

Toynbee's Industrial Revolution
Notes and Jottings (p. 243)

Before the close of the seventeenth century our forefathers consciously took their treasure out of religion and reinvested it in natural science.

The New York Times Magazine
A Turning Point in Man's Destiny (p. 5)
December 26, 1954

Valéry, Paul
Without religions the sciences would never have existed. For the human brain would not have trained itself to range beyond the immediate, ever-present "facts" of appearance which, for it, constitute reality.

The Collected Works of Paul Valéry
Volume 14
Analects
XLVIII (p. 285)

Virchow, Rudolf

There can be no scientific dispute with respect to faith, for science and faith exclude one another.

Disease, Life, and Man
On Man (p. 68)

. . . belief has no place as far as science reaches, and may be first permitted to take root where science stops.

Disease, Life, and Man
On Man (p. 69)

Whitehead, Alfred North

Religion will not gain its old power until it can face change in the same spirit as does science. Its principles may be eternal, but the expression of these principles requires continual development.

Science and the Modern World
Chapter XII (p. 189)

When we consider what religion is for mankind, and what science is, it is no exaggeration to say that the future course of history depends upon the decision of this generation as to the relations between them.

Science and the Modern World
Chapter XII (p. 181)

Science suggests a cosmology; and whatever suggests a cosmology suggests a religion.

Religion in the Making
Truth and Criticism (p. 126)

Wilde, Oscar

Science is the record of dead religions.

Phrases and Philosophies for the Use of the Young

SCIENCE AND WOMEN

de Lamennais, Felicite Robert

I have never met a woman who was competent to follow a course of reasoning the half of a quarter of an hour—*un demi quart d'heure*. She has qualities which are wanting in us, qualities of a particular, inexpressible charm; but, in the matter of reason, logic, the power to connect ideas, to enchain principles of knowledge and perceive their relationships, woman, even the most highly gifted, rarely attains to the height of a man of mediocre capacity.

In H.J. Mozans
Women in Science
Chapter III (p. 136)

Kant, Immanuel

All abstract speculations, all knowledge which is dry, however useful it may be, must be abandoned to the laborious and solid mind of man . . . For this reason women will never learn geometry.

In H.J. Mozans
Women in Science
Chapter III (p. 136)

Kass-Simon, G.
Farnes, Patricia

For women in science to be remembered, not only must their work be thought right, but usually it must have such impact upon scientific thought that exclusion is impossible. If women scientists are wrong, or if they narrowly miss the mark, or if they propound ideas that are ultimately superseded, not only are their ideas quickly forgotten, but as often as not, the women are ostracized by their contemporaries or treated with derision.

Women of Science
Introduction (p. xiii)

Lamy, M. Étienne

Women . . . group themselves at the center of human knowledge, whereas men disperse themselves toward its outer boundaries. While men are always pushing analysis to its utmost limits, women are seeking a synthesis. While men are becoming more technical, women are becoming more intellectual. They are better placed to observe the correlations of the different sciences, and to subordinate them to the common and unique source of truth from which they all descend. We seem, indeed, to be approaching a time when women will become the conservers of general ideas.

In H.J. Mozans
Women in Science
Chapter XII (pp. 409–10)

Mozans, H.J.

Whilst men of science will be forced to continue as specialists as long as the love of fame, to consider no other motives of research, continue to be a potent influence in their investigations, it is probable that women will have less love for the long and tedious processes involved in the more difficult kinds of specialization. They will, it seems likely, be more inclined to acquire a general knowledge of the whole circle of the sciences—a knowledge that will enable them to take a comprehensive survey of nature. And it will be fortunate for themselves, as well as for the men who must perforce remain specialists, if they elect to do so. For nothing gives falser views of nature as a whole, nothing more unfits the mind for a proper apprehension of higher and more important truths, nothing more incapacitates one for the enjoyment of the masterpieces of literature or the sweeter amenities of life, than the narrow occupation of a specialist who sees nothing in the universe but electrons, microbes and protozoa.

Women in Science
Chapter XII (pp. 408–9)

Myrdal, Sigrid

There's the question of how you react when your data do not turn out the way you want them to. One possibility is to think "Oh no, something went wrong, my experiment failed." or "Did I ask the question wrong?" and put the data in the drawer. I think the feminine approach is to ask "What's this trying to tell me?" and consider that nature may be more interesting and complicated than I expected, but therefore probably a bit more elegant. By actually having to deal with the data, I've gone to totally different interpretations. If something turns out quite screwy, I give it a chance. It's possible that it's more feminine to give something a chance.

In Linda Jean Shepherd
Lifting the Veil
Receptivity (p. 86)

Pizan, Christine de

I will tell you again—and don't fear a contradiction—if it were customary to send daughters to school like sons, and if they were then taught the natural sciences, they would learn as thoroughly and understand the subtleties of all the arts and sciences as well as sons.

The Book of the City of Ladies
I.27.1 (p. 63)

Plato

Nothing can be more absurd than the practice, which prevails in our country, of men and women not following the same pursuits with all their strength and with one mind, for thus the state, instead of being a whole, is reduced to a half.

In United Nations
Women and Science (p. 2)

Poullain de la Barre, François

L'esprit n'a point de sexe.
[The mind has no sex.]

De l'éducation des dames pour la conduite de l'esprit dans les sciences et dans les moeurs

Yentsch, Clarice M.
Sindermann, Carl J.

Science, as a remarkably conservative human institution despite its relatively brief history, has typically cast women in supporting roles in which they were subservient to male professionals, usually dreadfully underpaid, and totally unrecognized.

The Woman Scientist
Chapter 2 (p. 27)

SCIENTIFIC

Agassiz, Luis

Scientific investigation in our day should be inspired by a purpose as animating to the general sympathy, as was the religious zeal which built the Cathedral of Cologne or the Basilica of St. Peter's. The time is passed when men expressed their deepest convictions by these wonderful and beautiful religious edifices; but it is my hope to see, with the progress of intellectual culture, a structure arise among us which may be a temple of the revelations written in the material universe.

Louis Agassiz: His Life and Correspondence
Volume II
Dredging Expedition (pp. 670–1)

Bayliss, William M.

It is not going too far to say that the greatness of a scientific investigator does not rest on the fact of his having never made a mistake, but rather on his readiness to admit that he has done so, whenever the contrary evidence is cogent enough.

Principles of General Physiology
Preface (pp. xvi–xvii)

Boycott, A.E.

The difficulty in most scientific work lies in framing the questions rather than in finding the answers.

Nature
The Transition from Live to Dead (p. 93)
Volume 123, January 19, 1929

Butterfield, Herbert

[The scientific revolution] outshines everything since the rise of Christianity and reduces the Renaissance and Reformation to the ranks of mere episodes, mere internal displacements, within the system of medieval Christendom . . . it looms so large as the real origin of the modern world and the modern mentality that our customary periodization of European history has become an anachronism and an encumbrance.

The Origins of Modern Science
Introduction (pp. vii, viii)

Chargaff, Erwin

The scientific professions began to develop a momentum of their own, thereby creating a vested interest in always having more science, bigger science, better-endowed science. This is, incidentally, quite in contrast, for instance, to orchestra musicians whose influence on the number of orchestra pieces being written is minimal.

Perspectives in Biology and Medicine
Voices in the Labyrinth (p. 324)
Volume 18, Number 3, Spring 1975

As for scientific fashions, I should think that they last longer than women's fashions but less long than men's.

Perspectives in Biology and Medicine
Triviality in Science: A Brief Meditation on Fashions (p. 330)
Volume 19, Number 3, Spring 1976

Chesterton, G.K.

. . . the ordinary scientific man is strictly a sentimentalist. He is a sentimentalist in this essential sense, that he is soaked and swept away by mere associations.

Orthodoxy
The Ethics of Elfland (pp. 93–4)

Scientific phrases are used like scientific wheels and piston-rods to make swifter and smoother yet the path of the comfortable.

Orthodoxy
The Romance of Orthodoxy (p. 230)

Clifford, William Kingdon

Remember, then, that [scientific thought] is the guide of action; that the truth at which it arrives is not that which we can ideally contemplate without error, but that which we may act upon without fear; and you cannot fail to see that scientific thought is not an accompaniment of human progress, but human progress itself.

The Common Sense of the Exact Sciences (p. ii)

Compton, Karl Taylor

The geographical pioneer is now supplanted by the scientific pioneer . . . Without the scientific pioneer our civilization would stand still and our spirit would stagnate; with him mankind will continue to work toward his higher density. This being so, our problem is to make science as effective an element as possible in our American program for social progress.

A Scientist Speaks (p. 2)

Science requires straight and independent thinking. Every hypothesis or idea is capable of definite proof or disproof. The habit of mind that subjects every idea to rigid test is of utmost value. Much of the loose thinking in social, educational, political, and economic affairs would be avoided if the workers in these fields could be given a real training in accurate scientific thinking.

A Scientist Speaks (p. 39)

Cooper, Leon

I like to say sometimes that scientific fashion is like fashion in men's and women's clothes . . . One year the ties are wide; the next year they're narrow. One year the skirts are high; the next year they're low. And if everyone is wearing a short skirt, you're just hopelessly out of fashion if you're wearing a long skirt. That's the way it sometimes seems with science. You want to be in the middle of what everyone is talking about; you want to be in the mainstream. And the next year it might be something completely different.

In George Johnson
In the Palaces of Memory
A Model of Memory (p. 149)

Deason, Hilary J.

Scientific literacy has become a real and urgent matter for the informed citizen. Many have allowed themselves to lapse into a coma of scientific illiteracy because of the misconception that science and mathematics are beyond their understanding, have no personal appeal, and can be rejected or ignored. For them a tocsin has been sounded by hundreds of intelligent men and women whose personal crusade is the awakening of people everywhere to scientific awareness. They admonish, "read, mark, learn, and inwardly digest." They have spread an intellectual feast so rich and varied that scientific illiterates find that sharing in it is an exciting experience.

A Guide to Science Reading (p. ix)

Dunne, Finley Peter

There's always wan encouragin' thing about th' sad scientific facts that come out ivry week in th' pa-apers. They're usually not thrue.

Mr. Dooley On Making a Will
On the Descent of Man (p. 90)

Eddy, Mary Baker

Jesus of Nazareth was the most scientific man that ever trod the globe. He plunged beneath the material surface of things, and found the spiritual cause.

Science and Health with Key to the Scriptures
Chapter 10 (p. 313)

Editorial

The scientific illiteracy of politicians, their simple lack of 'feel' for what science is and what it can do, prevents them from exploring the deeper questions, among the most important facing humankind: how can science be conducted so that, on the one hand, the thinkers have the freedom to think, for that is the *sine qua non*; and how, on the other hand, can the products of unfettered thought be harnessed for the needs of society?

New Scientist
Who Cares About Science? (p. 18)
October 17, 1985

Einstein, Albert

To be sure, it is not the fruits of scientific research that elevate a man and enrich his nature, but the urge to understand, the intellectual work, creative or receptive.

Ideas and Opinions (p. 12)

Fischer, Martin H.

You must learn to talk clearly. The jargon of scientific terminology which rolls off your tongues is mental garbage.

In Howard Fabing and Ray Marr
Fischerisms

Fisher, R.A.

The concept that the scientific worker can regard himself as an inert item in a vast co-operative concern working according to accepted rules, is encouraged by directing attention away from his duty to form correct scientific conclusions, to summarize them and to communicate them to his scientific colleagues, and by stressing his supposed duty mechanically to make a succession of automatic "decisions" . . . The idea that this responsibility can be delegated to a giant computer programmed with Decision Functions belongs to the phantasy of circles rather remote from scientific research. The view has, however, really been advanced (Neyman, 1938) that Inductive Reasoning does not exist, but only "Inductive Behaviour"!

Statistical Methods and Scientific Inference (pp. 101–2)

Flexner, Abraham

So long as men strive to transcend their native powers, to rid themselves of prejudice and preconception, to observe phenomena in a dry light, the effort is scientific, whether at the moment it attains mathematical accuracy or not.

Medical Education
Chapter I (p. 3)

France, Anatole

. . . the scientific reasons for preferring one piece of evidence to another are sometimes very strong, but they are never strong enough to outweigh our passions, our prejudices, our interests, or to overcome that levity of mind common to all grave men. It follows that we continually present the facts in a prejudiced or frivolous manner.

Penguin Island
Preface (p. 4)

Goethe, Johann Wolfgang von

A scientific researcher must always think of himself as a member of a jury. His only concern should be the adequacy of the evidence and the clarity of the proofs which support it. Guided by this, he will form his opinion and cast his vote without regard for whether he shares the author's views.

Scientific Studies
Volume 12
Chapter VIII (pp. 306–7)

Groen, Janny
Smit, Eefke
Eijsvoogel, Juurd

Scientific information is essential, not only for the scientist. The politician, the entrepreneur and the public at large need to know about it too. The people in business find that neither the mass media nor the specialized scientific press are providing the information needed. General information is no longer enough, specialist information is only digestible for the learned. Who will bridge the gap?

The Discipline of Curiosity
Introduction (p. 4)

Huxley, Thomas

The scientific imagination always restrains itself within the limits of probability.

Collected Essays
Volume V
Science and Christian Tradition (p. 124)

. . . the scientific spirit is of more value than its products, and irrationally held truths may be more harmful than reasoned errors. Now the essence of the scientific spirit is criticism. It tells us that whenever a doctrine claims our assent we should reply, "Take it if you can compel it." The struggle for existence holds as much in the intellectual as in the physical world. A theory is a species of thinking, and its right to exist is coextensive with its power of resisting extinction by its rivals.

Collected Essays
Volume II
Darwiniana
The Coming of Age of "The Origin of Species" (p. 229)

King, Martin Luther, Jr.
The means by which we live have outdistanced the ends for which we live. Our scientific power has outrun our spiritual power. We have guided missiles and misguided men.

Strength to Love
Chapter VII (p. 57)

Latour, Bruno
Woolgar, S.
. . . our discussion is informed by the conviction that a body of practices widely regarded by outsiders as well organized, logical, and coherent, in fact consists of a disordered array of observations with which scientists struggle to produce order . . . Despite participants' well-ordered reconstructions and rationalizations, actual scientific practice entails the confrontation and negotiation of utter confusion.

Laboratory Life: The Social Construction of Scientific Facts (p. 36)

Lévi-Strauss, Claude
The scientific mind does not so much provide the right answers as ask the right questions.

The Raw and the Cooked
Overture (p. 7)

Lorand, Arnold
. . . we have often observed in persons whose lives have been devoted to serious scientific work, which has entirely absorbed them, a total absence of sexual desire for a long time, and even impotence.

Old Age Deferred
Chapter XLIX (p. 399)

Maxwell, James Clerk
One of the severest tests of a scientific mind is to discern the limits of the legitimate application of the scientific method.

The Scientific Papers of James Clerk Maxwell
Volume I (p. 759)

Medawar, Sir Peter

The purpose of scientific enquiry is not to compile an inventory of factual information, nor to build up a totalitarian world picture of Natural Laws in which every event that is not compulsory is forbidden. We should think of it rather as a logically articulated structure of justifiable beliefs about nature. It begins as a story about a Possible World—a story which we invent and criticize and modify as we go along, so that it winds up by being, as nearly as we can make it, a story about real life.

Pluto's Republic
Mainly About Intuition, Section 4 (pp. 110–1)

Oppenheimer, J. Robert

In any science there is harmony between practitioners. A man may work as an individual, learning of what his colleagues do through reading or conversation; he may be working as a member of a group on problems whose technical equipment is too massive for individual effort. But whether he is part of a team or solitary in his own study, he, as a professional, is a member of a community. His colleagues in his own branch of science will be grateful to him for the inventive or creative thoughts he has, will welcome his criticism . . . His world and work will be objectively communicable; and he will be quite sure that if there is error in it, that error will not long be undetected. In his own line of work he lives in a community where common understanding combines with common purpose and interest to bind men together both in freedom and in cooperation.

The Open Mind
Prospects in the Arts and Sciences (pp. 137–8)

Paulos, John Allan

In general, almost any mathematically expressed scientific fact can be transformed into a consumer caveat (or lure) that will terrify (or attract) people.

A Mathematician Reads the Newspaper
Asbestos Removal Closes NYC Schools (p. 142)

Payne-Gaposchkin, Cecilia

Do not undertake a scientific career in quest of fame or money. There are easier and better ways to reach them. Undertake it only if nothing else will satisfy you; for nothing else is probably what you will receive. Your reward will be the widening of the horizon as you climb. And if you achieve that reward you will ask no other.

An Autobiography and Other Recollections
Chapter 22 (p. 227)

Pearson, Karl

The scientific man has above all things to strive at self-elimination in his judgments . . .

The Grammar of Science
Introductory
Section 2 (p. 11)

Peirce, Charles Sanders

The scientific spirit requires a man to be at all times ready to dump his whole cartload of beliefs, the moment experience is against them.

In Justus Buchler (Editor)
Philosophical Writings of Peirce
Chapter 4 (pp. 46–7)

The man of action has to believe, the inquirer has to doubt; the scientific investigator is both.

In J.B. Conant
Modern Science and Modern Man
Science and Spiritual Values (p. 103)

Planck, Max

An important scientific innovation rarely makes its way by gradually winning over and converting its opponents: it rarely happens that Saul becomes Paul. What does happen is that its opponents gradually die out and that the growing generation is familiarized with the idea from the beginning . . .

The Philosophy of Physics
Chapter III (p. 97)

Poincaré, Henri

For a superficial observer, scientific truth is beyond the possibility of doubt; the logic of science is infallible, and if the scientists are sometimes mistaken, this is only from their mistaking its rules.

The Foundations of Science
Science and Hypothesis (p. 27)

Popper, Karl R.

Scientific theories are not the digest of observations, but they are inventions—conjectures boldly put forward for trial, to be eliminated if they clashed with observations; with observations which were rarely accidental, but as a rule undertaken with the definite intention of testing a theory by obtaining, if possible, a decisive refutation.

Conjectures and Refutations
Chapter 1, Section IV (p. 46)

The old scientific ideal of *episteme*—of absolutely certain, demonstrable knowledge—has proved to be an idol. The demand for scientific objectivity makes it inevitable that every scientific statement must remain tentative for ever. It may indeed be corroborated, but every corroboration is relative to other statements which, again, are tentative. Only in our subjective experiences of conviction, in our subjective faith, can we be 'absolutely certain' . . .

The Logic of Scientific Discovery
Chapter X, Section 85 (p. 280)

Price, Don K.

. . . most scientists are prepared to work most of the time within the framework of ideas developed by their acknowledged leaders. In that sense, within any discipline, science is ruled by oligarchs who hold influence as long as their concepts and systems are accepted as the most successful strategy . . . Once in a great while, a rival system is proposed; then there can usually be no settlement of the issue by majority opinion. The metaphor of "scientific revolution" suggests the way in which the losing party is displaced from authority, discredited and its doctrines eliminated from textbooks.

The Scientific Estate
Chapter 6 (p. 172)

Richet, Charles

Scientific doubt is a first-class quality, but rather eliminates piquancy from controversy.

The Natural History of a Savant
Chapter III (p. 25)

Rothschild, Lord Nathaniel Mayer

It is sometimes said in justification of basic research, that chance observations made during such work, and their subsequent study may be just as important as those made during applied R & D. While there is some truth in this contention, the country's needs are not so trivial as to be left to the mercies of a form of scientific roulette, with many more than the conventional 37 numbers on which the ball may land.

A Framework for Government Research and Development (p. 3)

Russell, Bertrand

. . . the scientific attitude is in some degree unnatural to man; the majority of our opinions are wish-fulfulments, like dreams in the Freudian theory.

The Scientific Outlook
Chapter I (p. 16)

Shapley, Harlow

Perhaps the greatest satisfaction in reading of scientific exploits and participating, with active imagination, in the dull chores, the brave syntheses, the hard-won triumphs of scientific work, lies in the realization that ours is not an unrepeatable experience. Tomorrow night we can go out again among the distant stars. Again we can drop cautiously below the ocean surface to observe the unbelievable forms that inhabit those salty regions of high pressure and dim illumination. Again we can assemble the myriad of molecules into new combinations, weave them into magic carpets that take us into strange lands of beneficent drugs and of new fabrics and utensils designed to enrich the process of everyday living . . . We can return another day to these shores, and once more embark for travels over ancient or modern seas in quest of half-known lands—go forth as dauntless conquistadors, outfitted with maps and gear provided through the work of centuries of scientific adventures. But we have done enough for this day. We have much to dream about. Our appetites may have betrayed our ability to assimilate. The fare has been irresistibly palatable. It is time to disconnect the magic threads; time to wind up the spiral galaxies, roll up the Milky Way and lay it aside until tomorrow.

In Harlow Shapley, Samuel Rapport and Helen Wright (Editors)
A Treasury of Science
On Sharing in the Conquests of Science (p. 4)

Skolimowski, Henryk

We are the proud inheritors and perpetuators of the scientific tradition. But perhaps also the slaves of certain modes of thinking; subjects to a conceptual tyranny which we glorify, thus being perfect slaves—slaves who enjoy their imprisonment.

In A.J. Ayala (Editor)
Studies in the Philosophy of Biology
Problems of Rationality in Biology (p. 213)

Spencer, Herbert

The truth is, that those who have never entered upon scientific pursuits know not a tithe of the poetry by which they are surrounded.

Education
Chapter I (p. 72)

Scientific truths, of whatever order, are reached by eliminating perturbing or conflicting factors, and recognizing only fundamental factors.

The Data of Ethics
Chapter XV, Section 104 (p. 268)

Tate, Allen

Scientific approaches, because each has its own partial conventions momentarily arrogating to themselves the authority of total explanation, must invariably fail to see all the experience latent in the work.

The New Republic
Critical Responsibility (p. 340)
Volume 51, Number 663, August 17, 1927

Valéry, Paul

Each mind can regard itself as a laboratory in which processes peculiar to the individual are used for transforming a substance common to all.

The results obtained by certain individuals are a source of wonderment to others. Starting out with ordinary carbon, one man produces a diamond, by means of temperatures and pressures that others never dreamt of. "Why, it's only carbon!" they say, after analyzing it. But they don't know how to do what he did.

The Collected Works of Paul Valéry
Volume 14
Analects (p. 482)

Weber, Max

In science, each of us knows that what he has accomplished will be antiquated in ten, twenty, fifty years. That is the fate to which science is subjected; it is the very *meaning* of scientific work, to which it is devoted in a quite specific sense, as compared with other spheres of culture . . . Every scientific "fulfillment" raises new "questions"; it *asks* to be surpassed and outdated. Whoever wishes to serve science has to resign himself to this fact . . . We cannot work without hoping that others will advance further than we have.

In H.H. Gerth and C. Wright Mills (Editors)
From Max Weber
Science as a Vocation (p. 138)

Weil, Simone

A scientific conception of the world doesn't prevent one from observing what is socially fitting.

The Need of Roots
Part Three (p. 248)

Weisskopf, Victor F.

Some people maintain that scientific insight has eliminated the need for meaning. I do not agree. The scientific worldview established the notion that there is a sense and purpose in the development of the universe when it recognized the evolution from the primal explosion to matter, life, and humanity. In humans, nature begins to recognize itself.

The Joy of Insight
Chapter Fourteen (pp. 317–8)

Westfall, Richard S.

The Scientific Revolution was the most important 'event' in Western history, and a historical discipline that ignores it must have taken an unhappy step in the direction of antiquarianism. For good and for ill, science stands at the center of every dimension of modern life. It has shaped most of the categories in terms of which we think, and in the process has frequently subverted humanistic concepts that furnished the sinews of our civilization. Through its influence on technology, it has helped to lift the burden of poverty from much of the Western world, but in doing so has accelerated our exploitation of the world's finite resources until already, not so long after the birth of modern science, we fear with good cause their exhaustion. Through its transformation of medicine, science has removed the constant presence of illness and pain, but it has also produced toxic materials that poison the environment and weapons that threaten us with extinction . . . I am convinced that the list describes a large part of the reality of the late twentieth century and that nothing on it is thinkable without the Scientific Revolution of the sixteenth and seventeenth centuries . . . I have yet to see the work that presents, in one integrated argument, the full position I just sketched so briefly, the position that offers the ultimate justification for the inclusion of the history of science prominently in any academic course that presumes to explain the origins of the world in which we live.

In H. Floris Cohen
The Scientific Revolution (p. 5)

Whitehead, Alfred North

The old foundations of scientific thought are becoming unintelligible. Time, space, matter, material, ether, electricity, mechanism, organism, configuration, structure, pattern, function, all require reinterpretation. What is the sense of talking about a mechanical explanation when you do not know what you mean by mechanics? The truth is that science started its modern career by taking over ideas derived from the weakest side of the philosophies of Aristotle's successors. In some respects it was a happy choice. It enabled the knowledge of the seventeenth century to be formulated so far as physics and chemistry were concerned, with a completeness which lasted to the present time. But the progress of biology and psychology has probably been checked by the uncritical assumption of half-truths. If science is not to degenerate into a medley of *ad hoc* hypotheses, it must become philosophical and must enter upon a thorough criticism of its own foundations.

Science and the Modern World
Chapter I (pp. 16–7)

Wittgenstein, Ludwig

In the course of a scientific investigation we say all kinds of things; we make many utterances whose role in the investigation we do not understand. For it isn't as though everything we say has a conscious purpose; our tongues just keep going. Our thoughts run in established routines, we pass automatically from one thought to another according to the techniques we have learned. And now comes the time for us to survey what we have said. We have made a whole lot of movements that do not further our purpose, or that even impede it, and now we have to clarify our thought processes philosophically.

Culture and Value (p. 64e)

Ziman, John

The moment of truth for many young scientists comes when they first act as a referee for a scientific paper; having striven for years to get their own work published *against* the criticism of anonymous referees, they find themselves, by psychological role-reversal, on the other side of the fence. Thus do we eventually internalize the 'scientific attitude'.

Reliable Knowledge
Chapter 6 (fn 13, p. 132)

The community of those who are competent to contribute to, or criticize, scientific knowledge must not be closed; it must be larger, and more open, than the group of those who entirely accept a current consensus or orthodoxy. It is an essential element in the health of Science, or of a science, or of the sciences, that self-confirming, mutually validating circles be unable to close. Yet it is also essential that technical scientific discussion be not smothered in a cloud of ignorant prejudices and cranky speculations.

Public Knowledge
Chapter 4 (p. 64)

SCIENTIFIC GLOSSARY

It has long been known that. . .
I haven't bothered to look up the original reference.

While it has not been possible to provide definite answers to these questions . . .
The experiments didn't work out, but I figured I could at least get a publication out of it.

High purity . . .
Very high purity . . .
Extremely high purity . . .
Super purity . . .
Composition unknown except for the exaggerated claim of the suppliers.

. . . accidentally strained during mounting.
. . . dropped on the floor.

. . . handled with extreme care throughout the experiments.
. . . not dropped on the floor.

It is clear that much additional work will be required before a complete understanding . . .
I don't understand it.

Unfortunately, a quantitative theory to account for these effects has not been formulated . . .
Neither has anybody else.

It is hoped that this work will stimulate further work in the field.
This paper isn't very good, but neither is any of the others on this miserable subject.

The agreement with the predicted curve is . . .
. . . excellent.
Fair.
. . . good.
Poor.

. . . satisfactory.
Doubtful.
. . . fair.
Imaginary.

As good as could be expected considering the approximations made in the analysis.
Non-existent.

Of great theoretical and practical importance.
Interesting to me.

Three of the samples were chosen for detailed study.
The results on the others didn't make sense and were ignored.

These results will be reported at a later date.
I might possibly get round to this sometime.

Typical results are shown.
The best results are shown.

Although some detail has been lost in reproduction, it is clear from the original micrograph that . . .
It is impossible to tell from the micrograph.

It is suggested . . .; It may be believed . . .; It may be that . . .
I think.

The most reliable values are those of Jones.
He was a student of mine.

It is generally believed that . . .
A couple of other guys think so too.

It might be argued that . . .
I have such a good answer to this objection that I shall now raise it.

Correct to within an order of magnitude.
Wrong.

Definitions from C.D. Graham Jr.
Metal Progress
Volume 71, May 1957 (pp. 75–6)

SCIENTIFIC PAPERS

Arber, Agnes

A record of research should not resemble a casual pile of quarried stone; it should seem "not built, but born", as Vasari said in praise of a building.

The Mind and the Eye
Chapter V (p. 50)

Chargaff, Erwin

I should like to find a way of discouraging unnecessary publications, but I have not found a solution, save the radical one . . . that all scientific papers be published anonymously.

Perspectives in Biology and Medicine
In Praise of Smallness—How Can We Return to Small Science (p. 383)
Volume 23, Number 3, Spring 1980

Dubos, René

. . . a scientific paper should never try to make more than one point.

In B.D. Davis
Perspectives in Biology and Medicine
Two Perspectives (p. 38)
Volume 35, Number 1, Autumn 1991

Mayo, William J.

Reading papers is not for the purpose of showing how much we know and what we are doing, but is an opportunity to learn.

Proceedings of Staff Meetings, Mayo Clinic
The Value of the Weekly General Staff Meeting
Volume 10, January 30, 1935

Medawar, Sir Peter

. . . it is no use looking to scientific 'papers,' for they not merely conceal but actively misrepresent the reasoning that goes into the work they describe . . . Nor is it any use listening to accounts of what scientists say they do, for their opinions vary widely enough to accommodate almost any methodological hypothesis we may care to devise. Only unstudied evidence will do—and that means listening at a keyhole.

The Art of the Soluble
Hypothesis and Imagination (p. 151)

The importance of a problem should not be judged by the number of pages devoted to it.
Albert Einstein and Leopold Infeld – (See p. 149)

SCIENTIST

Artaud, Antonin

But how is one to make a scientist understand that there is something unalterably deranged about differential calculus, quantum theory, or the obscene and so inanely liturgical ordeals of the precession of the equinoxes—. . .

In Susan Sontag
Selected Writings
Part 33
Van Gogh, the Man Suicided by Society (p. 497)

Bacon, Francis

The empiricists are like the ant; they only collect and use. The rationalists resemble the spiders, who make cobwebs out of their own substance. The scientist is like the bee; it takes a middle course; it gathers material from the flowers, but adapts it by a power of its own.

Novum Organum
XCV

Barr, Amelia

Whatever the scientists may say, if we take the supernatural out of life, we leave only the unnatural.

All the Days of My Life
Chapter 26 (p. 477)

Beveridge, W.I.B.

The scientist who has an independent mind and is able to judge the evidence on its merits rather than in light of prevailing conceptions is the one most likely to be able to realize the potentialities in something really new.

The Art of Scientific Investigation
Chance (p. 35)

Brain, Lord Walter Russell

Scientists . . . meet one another to exchange ideas, to promote their own particular branch of science, or science in general, or because they are aware of its social implications. Nevertheless, such collective activities, important though they may be in themselves, play a small part in their lives. Scientists, though they must always be aware of the work of their fellows in their own fields, are essentially individualists; and the body of knowledge to which they are contributing is an impersonal one. Apart from contributing to it, they have no collective consciousness, interest, or aim.

Science
Science and Antiscience (p. 193)
Volume 148, Number 3667, April 1965

Bronowski, Jacob

The most remarkable discovery made by scientists is science itself.

A Sense of the Future
Chapter 2 (p. 6)

Dissent is the native activity of the scientist, and it has got him into a good deal of trouble in the last years. But if that is cut off, what is left will not be a scientist. And I doubt whether it will be a man.

Science and Human Values
The Sense of Human Dignity (p. 61)

The society of scientists must be a democracy. It can keep alive and grow only by a constant tension between dissent and respect; between independence from the view of others, and tolerance from them.

Science and Human Values
The Sense of Human Dignity (pp. 62–3)

There never was a scientist who did not make bold guesses, and there never was a bold man whose guesses were not sometimes wild.

Science and Human Values
The Sense of Human Dignity (p. 64)

Burroughs, William

Too many scientists seem to be ignorant of the most rudimentary spiritual concepts. They tend to be suspicious, bristly, paranoid-type people with huge egos they push around like some elephantiasis victim with his distended testicles in a wheelbarrow terrified no doubt that some skulking ingrate of a clone student will sneak into his very brain and steal his genius work.

The Adding Machine
Immortality (p. 132)

Butler, Samuel

[Scientists] There are two classes, those who want to know, and do not care whether others think they know or not, and those who do not much care about knowing, but care very greatly about being reputed as knowing.

In Geoffrey Keynes and Brian Hill (Editors)
Samuel Butler's Notebooks
Scientists (p. 119)

Chargaff, Erwin

. . . outside his own ever-narrowing field of specialization, a scientist is a layman. What members of an academy of science have in common is a certain form of semiparasitic living.

Perspectives in Biology and Medicine
Bitter Fruits from the Tree of Knowledge
Section III (p. 492)
Volume 16, Number 4, Summer 1973

Great scientists are particularly worth listening to when they speak about something of which they know little; in their own specialty they are usually great and dull.

Heraclitean Fire (p. 85)

Chesterton, G.K.

Apparently a scientist is a man who surveys all the sciences, without any particular study of them, and then gives expression to his own moral principles or prejudices.

All is Grist
On Mr. Mencken and Fundamentalism (p. 50)

Clarke, Arthur C.

When a distinguished but elderly scientist states that something is possible, he is almost certainly right. When he states that something is impossible, he is very probably wrong.

Profiles of the Future
Chapter 2 (p. 14)

. . . scientists of over fifty are good for nothing except board meetings and should at all costs be kept out of the laboratory!

Profiles of the Future
Cahpter 2 (pp. 14–5)

Conant, James Bryant

. . . scientists today represent the progeny of one line of descent who migrated, so to speak, some centuries ago into certain fields which were ripe for cultivation. Once science had become self-propagating, those who till these fields have had a relatively easy time keeping up the tradition of their forebears.

Science and Common Sense
Chapter One (p. 13)

Cousteau, Jacques-Yves
What is a scientist after all? It is a curious man looking through a keyhole, the keyhole of nature, trying to know what's going on.

Christian Science Monitor
July 21, 1971

Cramer, F.
In the long run it pays the scientist to be honest, not only by not making false statements, but by giving full expression to facts that are opposed to his views. Moral slovenliness is visited with far severer penalties in the scientific than in the business world.

In W.I.B. Beveridge
The Art of Scientific Investigation
Scientists (p. 142)

Cronenberg, David
. . . everybody's a mad scientist, and life is their lab. We're all trying to experiment to find a way to live, to solve problems, to fend off madness and chaos.

In Chris Rodley
Cronenberg on Cronenberg
Chapter 1 (p. 7)

de Madariaga, Salvador
There are two kinds of scientists: they were once described . . . as the "why" and the "how". The how-scientist is mainly interested in the way things happen; the why-scientist seeks to find out the cause of things. The first is more of a technician; the second, more of a philosopher. The first is more of a man of talent; the second, more of a man of genius.

Essays with a Purpose
Science and Freedom (p. 43)

Dubos, René
Scientists, like artists, unavoidably reflect the characteristics of the civilization and the time in which they live. In this sense, they are "enchained" . . . by the inexorable logic of their time and their work.

Louis Pasteur, Free Lance of Science
Introduction (p. xxxviii)

. . . like other men, scientists become deaf and blind to any argument or evidence that does not fit into the thought pattern which circumstances have led them to follow.

Louis Pasteur, Free Lance of Science
Chapter VII (p. 197)

Dyson, Freeman
When something ceases to be mysterious it ceases to be of absorbing concern to scientists. Almost all the things scientists think and dream about are mysterious.

Infinite in All Directions
Chapter 2 (p. 14)

Egler, Frank E.
Scientists are only men, and are subject to all the foibles of their kind. They have the same drives for freedom, security, certainty, image and status as have other men. They have the same attraction for the known, the familiar and the comfortable, and will cling to old and sterile ideas like a broody hen sitting on boiled eggs. Like those others, there is a lunatic fringe, and a reasonable quota of social misfits, small-pool big-frogs, megalomaniacs, prima donnas, nymphomaniacs, gold diggers, entrepreneurs, prophets and devout disciples.

The Way of Science
The Nature of Science (p. 1)

Einstein, Albert
For the scientist, there is only "being," but no wishing, no valuing, no good, no evil—in short, no goal. As long as we remain within the realm of science proper, we can never encounter a sentence of the type: "Thou shalt not lie."

In Philipp Frank
Relativity—A Richer Truth
The Laws of Science and the Laws of Ethics (p. 9)

After a certain high level of technical skill is achieved, science and art tend to coalesce in esthetics, plasticity, and form. The greatest scientists are always artists as well.

In Alice Calaprice (Editor)
The Quotable Einstein (p. 171)

Eysenck, H.J.
Scientists, especially when they leave the particular field in which they have specialized, are just as ordinary, pig-headed and unreasonable as anybody else.

Omni
Continuum (p. 49)
Volume 2, December 1979

Faulkner, William

Our privacy . . . has been slowly and steadily and increasingly invaded until now our very dream of civilization is in danger. Who will save us but the scientist and the humanitarian? Yes, the humanitarian in science, and the scientist in the humanity of man.

In Warren Weaver
Science
Presidential Address
December 30, 1955

Feibleman, James K.

It is not the business of scientists to investigate just what the business of science is.

Technology and Culture
Pure Science, Applied Science, Technology, Engineering:
An Attempt at Definitions (p. 305)
Volume II, Number 4, Fall 1961

Feyerabend, Paul

Scientists are sculptors of reality—but sculptors in a special sense. They do not merely *act causally* upon the world (though they do that, too, and they have to if they want to "discover" new entities); they also *create semantic conditions* engendering strong inferences from known effects to novel projections and, conversely, from the projections to testable effects.

The Journal of Philosophy
Realism and the Historicity of Knowledge (pp. 404–5)
Volume LXXXVI, Number 8, 1989

Feynman, Richard P.

When a scientist doesn't know the answer to a problem, he is ignorant. When he has a hunch as to what the result is, he is uncertain. And when he is pretty darn sure of what the result is going to be, he is still in some doubt.

What Do You Care What Other People Think?
The Value of Science (p. 245)

It is our responsibility as scientists, knowing the great progress which comes from a satisfactory philosophy of ignorance, the great progress which is the fruit of freedom of thought, to proclaim the value of this freedom; to teach how doubt is not to be feared but welcomed and discussed; and to demand this freedom as our duty to all coming generations.

What Do You Care What Other People Think?
The Value of Science (p. 248)

Finniston, Sir Monty
You mustn't think scientists are stupid.

Observer
Sayings of the Week
January 16, 1983

Fitzgerald, Penelope
If they don't depend on true evidence, scientists are no better than gossips.

The Gate of Angels
Chapter 3 (p. 24)

Foster, A.O.
A true scientist is known by his confession of ignorance.

Quote
August 21, 1966 (p. 11)

Fuller, R. Buckminster
It has been customarily said by the public journals, assumedly bespeaking public opinion, that scientists "wrest order out of chaos." But the scientists who have made the great discoveries have been trying their best to tell the public that, as scientists, they have never found chaos to be anything other than the superficial confusion of innately *a priori* human ignorance at birth—an ignorance that is often burdened by the biases of others to remain gropingly unenlightened throughout its life. What the scientists have always found by physical experiment was an *a priori* orderliness of nature, or Universe always operating at an elegance level that made the discovering scientist's working hypotheses seem crude by comparison. The discovered reality made the scientists' exploratory work seem relatively disorderly.

In L.L. Larison Cudmore
The Center of Life (p. xi)

Goldenweiser, Alexander
The scientist, when in his laboratory, is craftsman and inventor in one. He also faces nature as a learner. Like the craftsman, he is prepared to commit errors and, having learned from them, to revise his procedure. Like the inventor, he is after something new, he plans his experiments deliberately, watches carefully, ever on the alert for a promising lead—a discovery.

Robots or Gods (p. 44)

. . . a scientist who is no longer capable of framing a hypothesis—or never was—is not a scientist but a methodological fossil.

Robots or Gods (p. 48)

Gornick, Vivian

Whatever a scientist is doing—reading, cooking, talking, playing—science thoughts are always there at the edge of the mind. They are the way the world is taken in; all that is seen is filtered through an everpresent scientific musing.

Women in Science
Part One (p. 39)

Harding, Rosamund E.M.

If the scientist has, during the whole of his life, observed carefully, trained himself to be on the look-out for analogy and possessed himself of relevant knowledge, then the 'instrument of feeling' . . . will become a powerful divining rod leading the scientist to discover order in the midst of chaos by providing him with a clue, a hint, or an hypothesis upon which to base his experiments.

An Anatomy of Inspiration
Chapter V (p. 86)

Hogan, James P.

Scientists are the easiest to fool . . . They think in straight, predictable, directable, and therefore misdirectable, lines. The only world they know is the one where everything has a logical explanation and things are what they appear to be. Children and conjurers—they terrify me. Scientists are no problem; against them I feel quite confident.

Code of the Lifemaker
Chapter I (p. 14)

Hubbard, Elbert

The scientist who now takes off his shoes knows that the place whereon he stands is holy ground. Science is reverent and speaks with lowered voice, for she has caught glimpses of mysteries undefinable, and to her have come thoughts that are beyond speech. Science cultivates the receptive heart and hospitable mind, and her prayer is for more light, and to that prayer the answer is even now coming.

In Albert Lane
Elbert Hubbard and His Work (p. 100)

Huggins, Charles

. . . there are two kinds of scientists—The "gee whiz" kind and the "so what" kind. Flies around the urine cause the first type to exclaim: "Gee whiz, what could that mean?" whereas the other says: "So what, let's clean up this mess and get on with a proper experiment."

In Elwood V. Jensen
Perspectives in Biology and Medicine
The Science of Science (p. 283)
Volume 12, Number 2, Winter 1969

Jensen, Elwood V.
Research among the less imaginative scientists has been likened to a fox-hunt. A creative investigator shouts "Tally-ho", and the entire troop rides off in the same direction.

Perspectives in Biology and Medicine
The Science of Science (p. 278)
Volume 12, Number 2, Winter 1969

Katscher, F.
That great scientists were believing Christians does not prove anything. In this century many free-thinkers have also made great contributions to science, scientific thinking and ethical questions regarding the application of science.

Nature
Correspondence (p. 390)
Volume 363, June 3, 1993

Koestler, Arthur
[Scientists are] Peeping Toms at the keyhole of eternity.

The Roots of Coincidence
Chapter 5, Section 9 (p. 140)

Kolbe, Monsignor
Two scientists are face to face with the visible world. One thinks of breaking it up to see what it is made of. Complex substances he analyzes into simple ones, largely ignoring what he may be losing in the analytical process. At last he comes to the irreducible elements, and is able to announce that all material things are composed of some ninety odd of them. Not satisfied with that, he goes on to break up even the atoms and finds that instead of the invisible, indestructible minima of last century, they are little 'solar systems,' consisting of positively electrified nuclei with one or two or many negative electrons dancing around them. Thus he looks upon the universe as made up of nuclei (of which he confesses he knows but little) and electrons (of which he claims to know that they are all exactly alike)—and nothing else. Thus he has totality divided into ultimate parts which can be mentally reassembled into the huge world-machine wound up ready to go.

The other scientist . . . takes full cognizance, with equal delight, of the skillful work of the analyst; but he says to him: You must have left something out, something which nature reveals. You take protoplasm from the dead body of a cat and from the dead body of a man, and you say that in both cases it is chemically a very complex substance, but that you can still give all the parts of it, and your analysis makes the two protoplasms the same. Something has been left out, for the life of man is much higher than the life of a cat, and in both cases it is the protoplasm that carries the life.

In Austin L. Porterfield
Creative Factors in Scientific Research (p. 229)

Kronberg, Arthur

A scientist . . . shouldn't be asked to judge the economic and moral value of his work. All we should ask the scientist to do is to find the truth—and then not keep it from anyone.

San Francisco Examiner
December 19, 1971

Kuhn, Thomas S.

Whether a scientist's work is predominantly theoretical or experimental, he normally seems to know, before his research project is even well under way, all the most intimate details of the result which that project will achieve. If the result is quickly forthcoming, well and good. If not, he will struggle with his apparatus and with his equations until, if at all possible, they yield results which conform to the sort of pattern which he has foreseen from the start.

The Structure of Scientific Revolutions (p. 348)

Larrabee, Eric

The only thing wrong with scientists is that they don't know where their own institution came from, what forces shaped and are still shaping it, and they have wedded themselves to an antihistorical way of thinking which threatens to deter them from ever finding out.

Commentary
Science and the Common People (p. 48)
June 1966

Leonard, Jonathan Norton

[A scientist's] real work is done in the silent hours of thought, the apparently aimless days of puttering around in the laboratory, and the mighty searching through reference books.

The World's Work
Steinmetz, Jove of Science, Part II (p. 140)
February 1929

Lewis, Gilbert N.

The scientist is a practical man and his are practical aims. He does not seek the *ultimate* but the *proximate*. He does not speak of the last analysis but rather of the next approximation . . . On the whole, he is satisfied with his work, for while science may never be wholly right it certainly is never wholly wrong; and it seems to be improving from decade to decade.

The Anatomy of Science
Chapter I (pp. 6–7)

Lewis, Sinclair

To be a scientist—it is not just a different job so that a man should choose between being a scientist and being an explorer or a bond-salesman or a physician or a king or a farmer. It is a tangle of ver-y obscure emotions, like mysticism, or wanting to write poetry; it makes its victim all different from the good natural man. The normal man, he does not care much what he does except that he should eat and sleep and make love. But the scientist is intensely religious—he is so religious that he will not accept quartertruths, because they are an insult to his faith.

Arrowsmith
Chapter XXVI, Section I (p. 279)

Mayo, Charles H.

The scientist is not content to stop at the obvious.

Collected Papers of the Mayo Clinic & Mayo Foundation
Problems in Medical Education
Volume 18, 1926

Medawar, Sir Peter

Ask a scientist what he conceives the scientific method to be and he will adopt an expression that is at once solemn and shifty-eyed: solemn, because he feels he ought to declare an option; shifty-eyed because he is wondering how to conceal the fact that he has no option to declare.

Pluto's Republic
Induction and Intuition in Scientific Thought
Part I, Section 2 (p. 80)

. . . scientists tend not to ask themselves questions until they can see the rudiments of an answer in their minds. Embarrassing questions tend to remain unasked or, if asked, to be answered rudely.

The Future of Man
Chapter 4 (p. 62)

The layman's interpretation of scientific practice contains two elements which seem to be unrelated and all but impossible to reconcile. In the one conception the scientist is a discoverer, an innovator, an adventurer into the domain of what is not yet known or not yet understood. Such a man must be speculative, surely, at least in the sense of being able to envisage what might happen or what could be true. In the other conception the scientist is a critical man, a skeptic, hard to satisfy; a questioner of received beliefs. Scientists (in this second view) are men of facts and not of fancies, and science is antithetical to, perhaps even an antidote to, imaginative activity in all its forms.

Induction and Intuition in Scientific Thought (p. 2)

Scientists are people of very dissimilar temperaments doing different things in very different ways. Among scientists are collectors, classifiers, and compulsive tidiers-up; many are detectives by temperament and many are explorers; some are artists and others artisans. There are poet-scientists and philosopher-scientists and even a few mystics.

The Art of the Soluble
Hypothesis and Imagination (p. 132)

Mencken, H.L.

[The Scientist] The value the world sets upon motives is often grossly unjust and inaccurate. Consider, for example, two of them: mere insatiable curiosity and the desire to do good. The latter is put high above the former, and yet it is the former that moves some of the greater men the human race has yet produced: the scientific investigators. What actually urges him on is not some brummagem idea of Service, but a boundless, almost pathological thirst to penetrate the unknown, to uncover the secret, to find out what has not been found out before. His prototype is not the liberator releasing slaves, the good Samaritan lifting up the fallen, but a dog sniffing tremendously at an infinite series of rat-holes.

Prejudices: Third Series (pp. 269–70)

Montessori, Maria

. . . what is a scientist? . . . We give the name scientist to the type of man who has felt experiment to be a means guiding him to search out the deep truth of life, to lift a veil from its fascinating secrets, and who, in this pursuit, has felt arising within him a love for the mysteries of nature, so passionate as to annihilate the thought of himself.

The Montessori Method
Chapter I (p. 8)

Movie

He was one of our greatest scientists. He has proved, beyond any question, that physical affection is purely electrochemical.

Silk Stockings
Cyd Charrise talking down love to Fred Astaire

Oppenheimer, J. Robert

It is proper to the role of the scientist that he not merely find new truth and communicate it to his fellows, but that he teach, that he try to bring the most honest and intelligible account of new knowledge to all who will try to learn.

Fifty Great Essays
Prospects in the Arts and Sciences

The true responsibility of a scientist, as we all know, is to the integrity and vigor of his science. And because most scientists like all men of learning, tend in part also to be teachers, they have a responsibility for the communication of the truths they have found. This is at least a collective, if not an individual responsibility. That we should see in this any insurance that the fruits of science will be used for man's benefit, or denied to man when they make for his distress or destruction, would be a tragic naiveté.

The Open Mind
Physics in the Contemporary World (p. 91)

Scientists aren't responsible for the facts that are in nature. It's their job to find the facts. There's no sin connected with it—no morals. If anyone should have a sense of sin, it's God. He put the facts there.

In Lincoln Barnett
Life
J. Robert Oppenheimer (p. 133)
October 10, 1949

Pasteur, Louis
When moving forward toward the discovery of the unknown, the scientist is like a traveler who reaches higher and higher summits from which he sees in the distance new countries to explore.

Quoted by René J. Dubos
Louis Pasteur, Free Lance of Science
Chapter III (p. 87)

Peabody, Francis
. . . the popular conception of the scientist as a man who works in a laboratory and who uses instruments of precision is as inaccurate as it is superficial, for a scientist is known, not by his technical processes, but by his intellectual processes; and the essence of the scientific method of thought is that it proceeds in an orderly manner toward the establishment of truth.

Journal of the American Medical Association
The Care of the Patient (p. 877)
Volume LXXXVIII, Number 12, March 19, 1927

Perelman, S.J.
I guess I'm just an old mad scientist at bottom. Give me an underground laboratory, half a dozen atomsmashers, and a beautiful girl in a diaphanous veil waiting to be turned into a chimpanzee, and I care not who writes the nation's laws.

Crazy Like a Fox
Captain Future, Block that Kick (p. 210)

Perry, Ralph Barton

Every scientist, furthermore, is himself a "self-made man." He owes his strictly scientific attainment to his own efforts and to the endowment with which nature has equipped him. Whatever elevation in life he reaches is not an artificial status created by institutions or traditions, but a measure of solid achievement. The scientist, therefore, respects man for what he is rather than for his class or station.

The Present Conflict of Ideals
Chapter IX (pp. 101–2)

Planck, Max

Since the real world, in the absolute sense of the word, is independent of individual personalities, and in fact of all human intelligence, every discovery made by an individual acquires a completely universal significance. This gives the inquirer, wrestling with his problem in quiet seclusion, the assurance that every discovery will win the unhesitating recognition of all experts throughout the entire world, and in this feeling of the importance of his work lies his happiness. It compensates him fully for many a sacrifice which he must make in his daily life.

Scientific Autobiography and Other Papers
The Meaning and Limits of Exact Science
Part III (p. 103)

Polanyi, M.

There are differences in rank between scientists, but these are of secondary importance: everyone's position is sovereign. The Republic of Science realizes the ideal of Rousseau, of a community in which each is an equal partner in a General Will. But this identification makes the General Will appear in a new light. It is seen to differ from any other will by the fact that it cannot alter its own purpose. It is shared by the whole community because each member of it shares in a joint task.

Science, Faith and Society
Background and Prospect (pp. 16–7)

Q.U.O.

What is a modern scientist? A modern scientist is, broadly speaking, a specialist of narrow outlook and, I fear, of narrow habit of mind. Broadly speaking, I say, for there are exceptions of importance but of limited number: persons of deep penetration and acquisitive powers; of open minds endowed with critical faculties. But clustering around these like the exterior of a nebula; or pendant from them like the tail of a comet, what do we find? Why, this. An indescribable mass of formless, shapeless satellites, almost invisible in their tenuosity, almost indistinguishable in their nebulosity; possessing minds of jelly-fish, revolving in orbits so circumscribed as to appear stationary. Who or what are these small souls, if souls they can be called? I cannot tell you who or what they are. But

I can tell you what they call themselves. They style themselves "We Scientists."

G.K.'s Weekly
We Scientists
February 22, 1930

Richards, I.A.

We believe a scientist because he can substantiate his remarks, not because he is eloquent and forcible in his enunciation. In fact, we distrust him when he seems to be influencing us by his manner.

Science and Poetry
Chapter II (p. 24)

Roszak, Theodore

. . . science rests itself not in the *world* the scientist beholds at any particular point in time, but in his mode of *viewing* that world. A man is a scientist not because of what he sees, but because of *how* he sees it.

The Making of a Counter Culture
Chapter VII (p. 213)

Ruse, Michael

A scientist should not cheat or falsify data or quote out of context or do any other thing that is intellectually dishonest. Of course, as always, some individuals fail; but science as a whole disapproves of such action. Indeed, when transgressors are detected, they are usually expelled from the community.

Science, Technology & Human Values
Response to the Commentary: *Pro Judice* (p. 74)
Volume 7, Number 41, Fall 1982

Rushton, J.P.

Research has suggested that scientists differ from non-scientists by exhibiting a high level of curiosity, especially at an early age, and in demonstrating a relatively low level of sociability. Scientists also tend to be shy, lonely, slow in social development, and indifferent to close personal relationships, group activities and politics. Other attributes include skepticism, preoccupation, reliability, and a facility for precise, critical thinking. Generally they are cognitively complex, independent, non-conformist, assertive, and unlikely to suppress thoughts and impulses; and, like successful entrepreneurs, eminent scientists are also calculated risk-takers.

Journal of Social and Biological Structure
Volume 11, 1980 (p. 140)

Seifriz, William

It is no matter of chance that the greatest scientists of all time, Copernicus, Newton, Kepler, Linnaeus, Faraday, Darwin, and Maxwell, were men of noble character, modest, straightforward, and full of human sympathy. The great French mathematician, Henri Poincaré, stated that the chief end of life is contemplation, not action.

Science
A New University (pp. 88–9)
Volume 120, Number 3107, July 16, 1954

Selye, Hans
The fairest thing we can experience is the mysterious. It is the fundamental emotion which stands at the cradle of true science. He who knows it not, and can no longer wonder, no longer feel amazement, is as good as dead. We all had this priceless talent when we were young. But as time goes by, many of us lose it. The true scientist never loses the faculty of amazement. It is the essence of his being.

Quoted in *Newsweek*
March 31, 1958

Scientists are probably the most individualistic bunch of people in the world. All of us are and should be essentially different; there would be no purpose in trying to fit us into a common mold.

From Dream to Discovery
Introduction

Smith, Homer W.
A scientist is one who, when he does not know the answer, is rigorously disciplined to speak up and say so unashamedly; which is the essential feature by which modern science is distinguished from primitive superstition, which knew all the answers except how to say, "I do not know."

From Fish to Philosopher
Chapter 13 (p. 210)

Snow, C.P.
Literary intellectuals at one pole—at the other scientists . . . Between the two a gulf of mutual incomprehension.

The Two Cultures: and A Second Look
Chapter I (pp. 11–2)

Twain, Mark
Scientists have odious manners, except when you prop up their theory; then you can borrow money off them.

What is Man? and Other Essays
The Bee (p. 283)

That is the way of the scientist. He will spend thirty years in building up a mountain range of facts with the intent to prove a certain theory; then he is so happy in his achievement that as a rule he overlooks the main chief fact of all—that his accumulation proves an entirely different thing.

What is Man? and Other Essays
The Bee (p. 283)

University of California, Berkeley
Scientists work better when they're all mixed-up.

Fortune
Advertisement insert after p. 8
April 14, 1986

Varèse, Edgard
Scientists are the poets of today.

Quoted in James Reagan
Artspace
Patrick Lysaght (p. 30)
Volume 9, Fall 1985

Walshe, Sir F.M.R.
It often is the cloistered scientist who knows least about men who is apt to pontificate most loudly and confidently about Man. Beware of him when he assures you that he knows all the answers about us, for too often his is one of those Peter Pans of science that every generation produces: a clever boy who hasn't grown up.

Canadian Medical Association Journal
Volume 67, 1962 (p. 395)

Weil, Simone
One could count on one's fingers the number of scientists throughout the world with a general idea of the history and development of their particular science: there is non who is really competent as regards sciences other than his own. As science forms an indivisible whole, one may say that there are no longer, strictly speaking, scientists, but only drudges doing scientific work.

Oppression and Liberty
Prospects (p. 13)

Weiss, Paul A.
Just like the painter, who steps periodically back from his canvas to gain perspective, so the laboratory scientist emerges above ground occasionally from the deep shaft of his specialized preoccupation to survey the cohesive, meaningful fabric developing from innumerable component tributary threads, spun underground much like his own. Only by such shuttling back and forth between the worm's eye view

of detail and the bird's eye view of the total scenery of science can the scientist gain and retain a sense of perspective and proportions.

In Arthur Koestler and J.R. Smythies
Beyond Reductionism (p. 3)

Whewell, William

We need very much a name to describe a cultivator of science in general. I should incline to call him a *Scientist*. Thus we might say, that as an Artist is a Musician, Painter, or Poet, a Scientist is a Mathematician, Physicist, or Naturalist.

The Philosophy of the Inductive Sciences Founded upon Their History (p. cxiii)

Whitehead, Alfred North

Many a scientist has patiently designed experiments for the *purpose* of substantiating his belief that animal operations are motivated by no purpose . . . Scientists animated by the purpose of proving that they are purposeless constitute an interesting subject for study.

The Function of Reason
Chapter I (p. 12)

A few generations ago the clergy, or to speak more accurately, large sections of the clergy were the standing examples of obscurantism. Today their place has been taken by scientists.

The Function of Reason
Chapter I (pp. 34–5)

Wiener, Norbert

. . . the degradation of the position of the scientist as independent worker and thinker to that of a morally irresponsible stooge in a science-factory has proceeded even more rapidly and devastatingly than I had expected.

Bulletin of the Atomic Scientists
A Rebellious Scientist After Two Years (p. 338)
Volume 4, Number 11, November 4, 1948

Wilder, Thornton

Then there is technology, the excess of scientists who learn how to make things much faster than we can learn what to do with them.

New York Times Magazine
In Plora Lewis
Thornton Wilder at 65 Looks Ahead—And Back (p. 28)
April 15, 1962

Wilson, Edward O.

Scientists do not discover in order to know, they know in order to discover.

Biophilia
The Poetic Species (p. 58)

The scientist is not a very romantic figure. Each day he goes into the laboratory or field energized by the hope of a great score. He is brother to the prospector and treasure hunter. Every little discovery is like a gold coin on the ocean floor. The professional's real business, the bone and muscle of the scientific endeavor, amounts to a sort of puttering: trying to find a good problem, thinking up experiments, mulling over data, arguing in the corridor with colleagues, and making guesses with the aid of coffee and chewed pencils until finally something—usually small—is uncovered. Then comes a flurry of letters and telephone calls, followed by the writing of a short paper in an acceptable jargon. The great majority of scientists are hard-working, pleasant journeymen, not excessively bright, making their way through a congenial occupation.

Biophilia
The Poetic Species (p. 59)

Ziman, John

A philosopher *is a person who knows less and less about more and more, until he knows nothing about everything.*

A scientist *is a person who knows more and more about less and less, until he knows everything about nothing.*

Knowing Everything About Nothing (p. v)

Zinsser, Hans

The scientist takes off from the manifold observations of predecessors, and shows his intelligence, if any, by his ability to discriminate between the important and the negligible, by selecting here and there the significant steppingstones that will lead across the difficulties to new understanding. The one who places the last stone and steps across to the terra firma of accomplished discovery gets all the credit. Only the initiated know and honor those whose patient integrity and devotion to exact observation have made the last step possible.

As I Remember Him
Chapter 20 (p. 332)

Zworykin, V.K.

Today, just about anything we can figure out on paper can be done in the laboratory, and eventually in the factory. Our technology has reached the stage where the scientist can safely say: "If we can write it down, we can do it."

The American Magazine
You Can Write It Down . . . (p. 123)
August 1949

SOLUTION

MacCready, Paul

When you do come up with a solution, you can always explain it logically, even though it's the absurd approach that gave you the solution.

In Kenneth A. Brown
Inventors at Work
Paul MacCready (p. 11)

Mencken, H.L.

. . . there is always a well-known solution to every human problem—neat, plausible, and wrong.

Prejudices: Second Series
The Divine Afflatus (p.158)

TECHNOLOGY

de Saint-Exupéry, Antoine

. . . the machine does not isolate man from the great problems of nature but plunges him more deeply into them.

Wind, Sand, and Stars
Chapter 3 (p. 67)

Drexler, K. Eric

People who confuse science with technology tend to become confused about limits . . . they imagine that new knowledge always means new know-how; some even imagine that knowing *everything* would let us do *anything*.

Engines of Creation (p. 148)

Embree, Alice

America's technology has turned in upon itself; its corporate form makes it the servant of profits, not the servant of human needs.

In Robin Morgan
Sisterhood is Powerful
Media Images 1 (p. 211)

Feynman, Richard P.

For a successful technology, reality must take precedence over public relations, for Nature cannot be fooled.

What Do You Care What Other People Think?
Appendix F (p. 237)

Mumford, Lewis

By his very success in inventing labor-saving devices, modern man has manufactured an abyss of boredom that only the privileged classes in earlier civilizations have ever fathomed.

The Conduct of Life
The Challenge of Renewal
Section 3 (p. 14)

Organisation for Economic Co-Operation and Development
Science and technology . . . have a number of distinguishing characteristics which cause special problems or complications. One . . . is ubiquity: they are everywhere. They are at the forefront of social change. They not only serve as agents of change, but provide the tools for analysing social change. They pose, therefore, special challenges to any society seeking to shape its own future and not just to react to change or to the sometimes undesired effects of change.

Technology on Trial: Public Participation in Decision-Making Related to Science and Technology
Chapter I, Section B (p. 16)

Thoreau, Henry David
Our inventions are wont to be pretty toys, which distract our attention from serious things. They are but improved means to an unimproved end.

Walden
Economy

Unknown
You ask me about the scientist and social responsibility. I have definite convictions here. A pure scientist must not deny himself a discovery by worrying about social consequences. He can't possibly know the practical derivatives of his work and, therefore, cannot be held responsible for the eventual use to which his discoveries may be put. Technology, however, is different. While not separable from science, it nevertheless affects society directly and must be managed by society.

In Marlan Blissett
Politics in Science
Chapter 3 (p. 59)

Wilkins, Maurice
Science, with technology, is the only way we have to avoid starvation, disease, and premature death. The misapplication of science and technology is due to the fact that the politics are wrong. Now my own view is that the politics are indeed wrong; but politics and science are so closely interrelated that they can hardly be separated.

In Horace Freeland Judson
The Eighth Day of Creation
DNA, You Know, Is Midas' Gold (p. 97)

TEMPLE OF SCIENCE

Einstein, Albert

In the temple of science are many mansions, and various indeed are they that dwell therein and the motives that have led them thither. Many take to science out of a joyful sense of superior intellectual power; science is their own special sport to which they look for vivid experience and the satisfaction of ambition; many others are to be found in the temple who have offered the products of their brains on this altar for purely utilitarian purposes. Were an angel of the Lord to come and drive all the people belonging to these two categories out of the temple, the assemblage would be seriously depleted, but there would still be some men, of both present and past times, left inside.

Ideas and Opinions
Principles of Research (p. 224)

Many kinds of men devote themselves to Science, and not all for the sake of Science herself. There are some who come into her temple because it offers them the opportunity to display their particular talents. To this class of men science is a kind of sport in the practice of which they exult, just as an athlete exults in the exercise of his muscular prowess. There is another class of men who come into the temple to make an offering of their brain pulp in the hope of securing a profitable return. The men are scientists only by the chance of some circumstance which offered itself when making a choice of career . . . it is clear that if the men who have devoted themselves to science consisted only of the two categories I have mentioned, the edifice could never have grown to its present proud dimensions . . . I am inclined to agree with Schopenhauer in thinking that one of the strongest motives that lead people to give their lives to art and science is the urge to flee from everyday life, with its drab and deadly dullness, and thus to unshackle the chains of one's own transient desires, which supplant one another in interminable succession so long as the mind is fixed on the horizon of daily environment.

In Max Planck
Where Is Science Going?
Prologue (p. 7)

Pasteur, Louis

Preconceived ideas are like searchlights which illuminate the path of the experimenter and serve him as a guide to interrogate nature. They become a danger only if he transforms them into fixed ideas—this is why I should like to see these profound words inscribed on the threshold of all the temples of science: 'The greatest derangement of the mind is to believe in something because one wishes it to be so' . . .

In René Dubos
Louis Pasteur, Free Lance of Science
Speech to the French Academy of Medicine
July 18, 1876
Chapter XIII (p. 376)

Planck, Max

Anybody who has been seriously engaged in scientific work of any kind realizes that over the entrance to the gates of the temple of science are written the words: *Ye must have faith*. It is a quality which the scientist cannot dispense with.

Where is Science Going?
Epilogue (p. 214)

THEORY

Boltzmann, Ludwig

A friend of mine has defined the practical man as one who understands nothing of theory and the theoretician as an enthusiast who understands nothing at all.

Theoretical Physics and Philosophical Problems
On the Significance of Theories (p. 33)

Cooper, Leon

A theory is a well-defined structure that we hope is in correspondence with what we observe. It's an architecture, a cathedral.

In George Johnson
In the Palaces of Memory
The Memory Machine (pp. 114–5)

Darwin, Charles

Let theory guide your observations, but till your reputation is well established, be sparing in publishing theory. It makes persons doubt your observations.

More Letters of Charles Darwin
Volume II
To Scott
June 6, 1863 (p. 323)

Duschl, R.

In order to say we have developed a knowledge of science, we must be able to say we have an understanding of the function, structure, and generation of scientific theories.

Restructuring Science Education
Chapter 6 (p. 96)

Einstein, Albert

It is theory which decides what we can observe.

In Werner Heisenberg
Physics and Beyond
Chapter 6 (p. 77)

The scientific theorist is not to be envied. For Nature, or more precisely experiment, is an inexorable and not very friendly judge of his work. It never says "Yes" to a theory. In the most favorable cases it says "Maybe," and in the great majority of cases simply "No." If an experiment agrees with a theory it means for the latter "Maybe," and if it does not agree it means "No." Probably every theory will some day experience its "No"—most theories, soon after conception.

Note dated November 11, 1922
Quoted in Helen Dukas and Banesh Hoffmann
Albert Einstein: The Human Side (p. 18)

Einstein, Albert
Infeld, Leopold

There are no eternal theories in science. It always happens that some of the facts predicted by a theory are disproved by experiment. Every theory has its period of gradual development and triumph, after which it may experience a rapid decline.

The Evolution of Physics
The Decline of the Mechanical View (p. 77)

Emerson, Ralph Waldo

We do what we can, and then make a theory to prove our performance the best.

Journals
1834

Faraday, Michael

The world little knows how many of the thoughts and theories which have passed through the mind of a scientific investigator have been crushed in silence and secrecy; that in the most successful instances not a tenth of the suggestions, the hopes, the wishes, the preliminary conclusions have been realized.

Quoted in W.I.B. Beveridge
The Art of Scientific Investigation
Imagination (p. 58)

Gay-Lussac, Joseph Louis

In order to draw any conclusion . . . it is prudent to wait until more numerous and exact observations have provided a solid foundation on which we may build a rigorous theory.

In Maurice Crosland
Gay-Lussac: Scientist and Bourgeois
Chapter 4 (p. 71)

Hazlitt, William

A favourite theory is a possession for life . . .

Characteristics
CXVII (p. 474)

Hubble, Edwin

No theory is sacred. When a theory fails to meet the test of verified predictions, it is modified to include the larger field, or, very rarely, it may be abandoned completely.

The Nature of Science
Experiment and Experience (p. 41)

Huxley, Thomas

The struggle for existence holds as much in the intellectual as in the physical world. A theory is a species of thinking, and its right to exist is coextensive with its power of resisting extinction by its rivals.

Collected Essays
Volume II
Darwiniana
Coming of Age of "The Origin of Species" (p. 229)

Lakatos, I.

No theory forbids some state of affairs specifiable in advance; it is not that we propose a theory and Nature may shout NO. Rather, we propose a maze of theories, and Nature may shout INCONSISTENT.

Proceedings of the Aristotelian Society
Criticism and the Methodology of Scientific Research Programmes (p. 162)
Volume 69, 1968–1969

Mach, Ernst

The object of natural science is the connexion of phenomena; but the theories are like dry leaves which fall away when they have long ceased to be the lungs of the tree of science.

History and Root of the Principle of the Conservation of Energy
Chapter IV (p. 74)

Medawar, Sir Peter

Scientific theories . . . begin as imaginative constructions. They begin, if you like, as stories, and the purpose of the critical or rectifying episode in scientific reasoning is precisely to find out whether or not these stories are about real life.

Pluto's Republic
Science and Literature
Section 4 (p. 53)

Poincaré, Henri

It is not sufficient for a theory to affirm no false relations, it must not hide true relations.

The Foundations of Science
Science and Hypothesis
The Theories of Modern Physics (p. 145)

Popper, Karl R.

. . . our critical examinations of our theories lead us to attempts to test and to overthrow them; and these lead us further to experiments and observations of a kind which nobody would ever have dreamed of without the stimulus and guidance both of our theories and of our criticisms of them. For indeed, the most interesting experiments and observations were carefully designed in order to test our theories, especially our new theories.

Conjectures and Refutations
Chapter 10, Section I (pp. 215–6)

The dogmatic attitude of sticking to a theory as long as possible is of considerable significance. Without it we could never find out what is in a theory—we should give the theory up before we had real opportunity of finding out its strength; and in consequence no theory would ever be able to play its role of bringing order into the world, of preparing us for future events, of drawing our attention to events we should otherwise never observe.

Conjectures and Refutations
Chapter 15 (fn 1, p. 312)

Theories are nets cast to catch what we call 'the world': to rationalize, to explain, and to master it. We endeavour to make the mesh ever finer and finer.

The Logic of Scientific Discovery
Chapter III (p. 59)

Every 'good' scientific theory is one which forbids certain things to happen; the more a theory forbids, the better it is.

In C.A. Mace (Editor)
British Philosophy in the Mid-Century
Philosophy of Science: A Personal Report
I (p. 159)

. . . in order that a new theory should constitute a discovery or a step forward it should conflict with its predecessor . . . it should contradict its predecessor; it should overthrow it. In this sense, progress in science—or at least a striking progress—is always revolutionary.

In Rom Harré
Problems of Scientific Revolution
The Rationality of Scientific Revolutions (pp. 82–3)

Richards, Dickinson W.

The problems are the ones that we have always known. The little gods are still with us, under different names. There is conformity: of technique, leading to repetition; of language, encouraging if not imposing conformity of thought. There is popularity: it is so easy to ride along on an already surging tide; to plant more seed in an already well-ploughed field; so hard to drive a new furrow into stony ground. There is laxness: the disregard of small errors, of deviations, of the unexpected response; the easy worship of the smooth curve. There is also fear: the fear of speculation; the overprotective fear of being wrong. We are forgetful of the curious and wayward dialectic of science, whereby a well-constructed theory even if it is wrong, can bring a signal advance.

Transactions of the Association of American Physicians
Volume 75, 1962 (p. I)

Richet, Charles

I often recall to my students the history of Don Quixote, who, having constructed a helmet of cardboard and wood, wished to prove its solidity. Alas, the poor helmet flew to bits when his own good sword struck it. Then the knight, no whit discouraged, made a new and stronger helmet. He raised his sword. "No," said he, "I will not smite it; my helmet might perhaps be broken." Let us not imitate Don Quixote. Let us not fear to submit our helmet (whether of theory or experiment) to two, three, or even six tests, and perhaps more.

The Natural History of a Savant
Chapter II (p. 17)

Russell, Bertrand

. . . it is only theory that makes men completely incautious.

Unpopular Essays
Ideas That Have Harmed Mankind (p. 163)

Santayana, George

Theory helps us to bear our ignorance of facts.

The Sense of Beauty
The Average Modified in the Direction of Pleasure (p. 125)

Schiller, F.C.S.

It is the business of theories to forecast 'facts', and of facts to form points of departure for theories, which again, when verified by the new facts to which they have successfully led, will extend the borders of knowledge.

In Charles Singer (Editor)
Studies in the History and Method of Science
Volume I
Scientific Discovery and Logical Proof (p. 275)

Silver, Brian L.

Facts may be regarded as indisputable; theories are not.

The Ascent of Science
Chapter 2 (p. 19)

Skolimowski, Henryk

Theories, like old soldiers, fade away rather than being killed on the scientific battlefield.

In A.J. Ayala (Editor)
Studies in the Philosophy of Biology
Problems of Rationality in Biology (p. 217)

Stevenson, Robert Louis

It is better to emit a scream in the shape of a theory than to be entirely insensible to the jars and incongruities of life and take everything as it comes in a forlorn stupidity.

Virginibus Puerisque
Crabbed Age and Youth (p. 102)

Thomson, J.J.

A scientific theory is a tool and not a creed.

In Richard Willstätter
From My Life
Chapter 12 (p. 388)

Toulmin, Stephen E.

It is part of the art of the sciences, which has to be picked up in the course of the scientist's training, to recognize the situations which any particular theory or principle can be applied to, and when it will cease to hold.

The Philosophy of Science
Chapter III (pp. 92–3)

Twain, Mark

. . . the trouble about arguments is, they ain't nothing but *theories*, after all, and theories don't prove nothing, they only give you a place to rest on, a spell, when you are tuckered out butting around and around trying to find out something there ain't no way to find out . . . There's another trouble about theories: there's always a hole in them somewheres, sure, if you look close enough.

The Complete Works of Mark Twain
Volume 14
Tom Sawyer Abroad
Chapter IX (p. 78)

van Fraassen, Bas C.

Science aims to give us, in its theories, a literally true story of what the world is like; and acceptance of a scientific theory involves the belief that it is true.

The Scientific Image (p. 8)

Waddington, C.H.

A scientific theory cannot remain a mere structure within the world of logic, but must have implications for action and that in two different ways. In the first place, it must involve the consequence that if you do so and so, such and such results will follow. That is to say it must give, or at least offer the possibility of controlling the process; and secondly—and this is a point not so often mentioned by those who discuss the nature of scientific theories—its value is quite dependent on its power of suggesting the next step in scientific advance.

The Nature of Life
Chapter I (pp. 11–2)

Wisdom, J.O.

[Theory] Sometimes it is used for a hypothesis, sometimes for a confirmed hypothesis; sometimes for a train of thought; sometimes for a wild guess at some fact or for a reasoned claim about what some fact is—or even for a philosophical speculation.

Foundations of Inference in Natural Science
Chapter III (p. 33)

Ziman, John

The verb 'to theorize' is now conjugated as follows: 'I built a model; you formulated a hypothesis; he made a conjecture.'

Reliable Knowledge
Chapter 2 (fn 20, p. 22)

TRUTH

Aronowitz, Stanley

The power of science consists, in the first place, in its conflation of knowledge and truth. Devising a method of proving the validity of propositions about objects taken as external to the knower has become identical with what we mean by truth.

Science as Power: Discourse and Ideology in Modern Society
Preface (p. vii)

Bacon, Francis

Truth emerges more readily from error than from confusion.

Novum Organum
In Ritchie Calder
Man and the Cosmos (p. 19)

Beaumont, William

Truth, like beauty, when "unadorned, is adorned the most;" and in prosecuting these experiments and inquiries, I believe I have been guided by its light.

Experiments and Observations on the Gastric Juice and the Physiology of Digestion
Section 5 (p. 260)

Bernard, Claude

It seems, indeed, a necessary weakness of our mind to be able to reach truth only across a multitude of errors and obstacles.

An Introduction to the Study of Experimental Medicine
Part III, Chapter I, Section ii (p. 170)

Bohr, Niels

The opposite of a correct statement is a false statement. But the opposite of a profound truth may well be another profound truth.

In Werner Heisenberg
Physics and Beyond
Chapter 8 (p. 102)

Bronowski, Jacob

We cannot define truth in science until we move from fact to law. And within the body of laws in turn, what impresses us as truth is the orderly coherence of the pieces. They fit together like the characters of a great novel, or like the words of a poem. Indeed, we should keep that last analogy by us always, for science is a language, and like a language it defines its parts by the way they make up meaning. Every word in a sentence has some uncertainty of definition, and yet the sentence defines its own meaning and that of its words conclusively. It is the internal unity and coherence of science which gives it truth, and which makes it a better system of prediction than any less orderly language.

The Common Sense of Science
Chapter VIII
Section 5 (p. 131)

Truth in science is like Everest, an ordering of the facts.

Science and Human Values
The Sense of Human Dignity (p. 52)

Chandler, Raymond

There are two kinds of truth: the truth that lights the way and the truth that warms the heart. The first of these is science, and the second is art . . . Without art science would be as useless as a pair of high forceps in the hands of a plumber. Without science art would become a crude mess of folklore and emotional quackery.

The Notebooks of Raymond Chandler
Great Thought (p. 7)

Chargaff, Erwin

The initial incommunicability of truth, scientific or otherwise, shows that we think in grooves, and that it is painful for us to be torn away from the womblike security of accepted concepts.

Heraclitean Fire (p. 86)

Dewey, John

There is but one sure road of access to truth—the road of coöperative inquiry operating by means of observation, experiment, record, and controlled reflection.

Common Faith
Chapter II (p. 32)

Einstein, Albert

It is difficult even to attach a precise meaning to the term "scientific truth". Thus the meaning of the word "truth" varies according to whether we deal with a fact of experience, a mathematical proposition, or a scientific theory. "Religious truth" conveys nothing clear to me at all.

Ideas and Opinions
On Scientific Truth (p. 261)

Truth is what stands the test of experience.

In Philipp Frank
Relativity—A Richer Truth
The Laws of Science and the Laws of Ethics (p. 10)

As for the search for truth, I know from my own painful searching, with its many blind alleys, how hard it is to take a reliable step, be it ever so small, towards the understanding of that which is truly significant.

Letter dated February 13, 1934
Quoted in Helen Dukas and Banesh Hoffmann
Albert Einstein: The Human Side (p. 18)

Esquivel, Laura

Anything could be true or false, depending on whether one believed it.

Like Water for Chocolate
July (p. 127)

Gray, George W.

No truth is sacrosanct. No belief is too generally accepted, too well established by experiment, to escape the challenge of doubt. And no doubt is too radical to receive a hearing if it is seriously proposed.

Harper's Monthly Magazine
The Riddle of Our Reddening Skies (p. 169)
July 1937

Heaviside, Oliver

We do not dwell in the Palace of Truth. But, as was mentioned to me not long since, "There is a time coming when all things shall be found out." I am not so sanguine myself, believing that the well in which Truth is said to reside is really a bottomless pit.

Electromagnetic Theory
Chapter I
Volume I (p. 1)

Heinlein, Robert A.

The hardest part about gaining any new idea is sweeping out the false idea occupying that niche. As long as that niche is occupied, evidence and proof and logical demonstration get nowhere. But once the niche is emptied of the wrong idea that has been filling it—once you can honestly say, 'I don't know,' then it becomes possible to get at the truth.

The Cat Who Walks Through Walls (p. 244)

Hoagland, H.

Science assumes that the quest for truth is a major end in itself, not only for the individual but for society as a whole, even though we know that ultimate, final truth with a capital T is not to be found.

Science
Science and the New Humanism (p. 112)
Volume 143, 1964

Huxley, Aldous

Science is the only way we have of shoving truth down the reluctant throat.

Literature and Science
Chapter 27 (p. 79)

Huxley, Thomas

Ecclesiasticism in science is only unfaithfulness to truth.

Collected Essays
Volume II
Darwiniana
Mr. Darwin's Critics (p. 149)

History warns us, however, that it is the customary fate of new truths to begin as heresies and to end as superstitions . . .

Collected Essays
Volume II
Darwiniana
The Coming of Age of "The Origin of Species" (p. 229)

Science has fulfilled her function when she has ascertained and enunciated truth . . .

Collected Essays
Volume VII
Man's Place in Nature
On the Relations of Man to the Lower Animals (p. 151)

Jonson, Ben

If in some things I dissent from others, whose wit, industry, diligence, and judgement, I look up at and admire, let me not therefore hear presently of ingratitude and rashness. For I thank those that have taught me, and ever will; but yet dare not think the scope of their labour and inquiry was to envy their posterity what they also could add and find out . . . If I err, pardon me . . .

Discoveries Made Upon Man and Matter (p. 19)

Lawson, Alfred William

Education is the science of knowing TRUTH.
Miseducation is the art of absorbing FALSITY.
TRUTH is that which is, not that which ain't.
FALSITY is that which ain't, not that which is.

Quoted by Martin Gardner in
Fads and Fallacies
Chapter 6 (p. 76)

Le Bon, Gustave

Science has promised us truth—an understanding of such relationships as our minds can grasp; it has never promised us either peace or happiness.

La Psychologie des Foules
Introduction

Lessing, Gotthold Ephraim

Not in the possession of truth but the effort in struggling to attain it brings joy to the researcher.

In Max Planck
Where Is Science Going?
From the Relative to the Absolute (p. 200)

Levy, Hyman

Truth is a dangerous word to incorporate within the vocabulary of science. It drags with it, in its train, ideas of permanence and immutability that are foreign to the spirit of a study that is essentially an historically changing movement, and that relies so much on practical examination within restricted circumstances . . . Truth is an absolute notion that science, which is not concerned with any such permanency, had better leave alone.

The Universe of Science
Chapter V (pp. 206–7)

Lewis, Gilbert N.

The theory that there is an ultimate truth, although very generally held by mankind, does not seem useful to science except in the sense of a horizon toward which we may proceed, rather than a point which may be reached.

The Anatomy of Science
Chapter I (p. 7)

Maxwell, James Clerk

For the sake of these different types, scientific truth should be presented in different forms, and should be regarded as equally scientific, whether it appears in the robust form of vivid colouring of a physical illustration, or in the tenuity and paleness of a symbolical expression.

The Collected Papers of James Clerk Maxwell
Volume 2
XLI
Address to the Mathematical and Physical Sections of the British Association
September 15, 1870 (p. 220)

Millikan, Arthur

. . . in science, truth once discovered always remains truth.

Science and the New Civilization
Chapter III (p. 76)

Newton, Sir Isaac

I do not know what I may appear to the world, but to myself I seem to have been only like a boy playing on the sea-shore, and diverting myself in now and then finding a smoother pebble or a prettier shell than ordinary, whilst the great ocean of truth lay all undiscovered before me.

In David Brewster
Memoirs of the Life, Writings and Discoveries of Sir Isaac Newton
Chapter 27 (p. 331)

Pasteur, Louis

Truth, Sir, is a great coquette. She will not be won by too much passion. Indifference is often more successful with her. She escapes when apparently caught, but she yields readily if patiently waited for. She reveals herself when one is about to abandon the hope of possessing her; but she is inexorable when one affirms her, that is when one loves her with too much fervor.

In René Dubos
Louis Pasteur, Free Lance of Science
Chapter XIV (p. 389)

Peirce, Charles Sanders

. . . truths, on the average, have a greater tendency to get believed than falsities have. Were it otherwise, considering that there are myriads of false hypotheses to account for any given phenomenon, against one sole true one (or if you will have it so, against every true one), the first step towards genuine knowledge must have been next door to a miracle.

The Collected Works of Charles Sanders Peirce
Volume 5
Pragmatism and Pragmaticism (p. 431)

Penrose, Roger

Scientists do not invent truth—they discover it.

In John Horgan
Scientific American
Quantum Consciousness (p. 32)
Volume 261, Number 5, November 1989

Planck, Max

A new scientific truth does not triumph by convincing its opponents and making them see the light, but rather because its opponents eventually die, and a new generation grows up that is familiar with it.

Scientific Autobiography and Other Papers
A Scientific Autobiography (pp. 33–4)

Popper, Karl R.

We have no reason to regard the new theory as better than the old theory—to believe that it is nearer to the truth—until we have derived from the new theory *new predictions* which were unobtainable from the old theory . . . and until we have found that these new predictions were successful.

Conjectures and Refutations
Chapter 10, Section XXI (p. 246)

Reichenbach, Hans

He who searches for truth must not appease his urge by giving himself up to the narcotic of belief. Science is its own master and recognizes no authority beyond its confines.

The Rise of Scientific Philosophy
Chapter 12 (p. 214)

Renan, Ernest

The simplest schoolboy is now familiar with truths for which Archimedes would have sacrificed his life.

Quoted by L.I. Ponomarev in
The Quantum Dice (p. 34)

Science has no enemies save those who consider truth as useless and making no difference, and those who granting to truth its priceless value profess to get at it by other roads than those of criticism and rational investigation.

The Future of Science
Chapter IV (p. 68)

Richet, Charles

. . . if you would discover a new truth, do not seek to know what use will be made of it.

The Natural History of a Savant
Chapter XII (p. 133)

Romanoff, Alexis

Science speaks the language of universal truth.

Encyclopedia of Thoughts
Aphorism 961

Spencer-Brown, George

To arrive at the simplest truth, as Newton knew and practiced, *requires* years of *contemplation*. Not activity. Not reasoning. Not calculating. Not busy behavior of any kind. Not reading. Not talking. Not making an effort. Not thinking. Simply *bearing in mind* what it is one needs to know. And yet those with the courage to tread this path to real discovery are not only offered practically no guidance on how to do so, they are actively discouraged and have to set about it in secret, pretending meanwhile to be diligently engaged in the frantic diversions and to conform with the deadening personal opinions which are continually being thrust upon them.

Laws of Form
Appendix I (p. 110)

Tennyson, Alfred Lord

Like truths of Science waiting to be caught.

The Complete Poetical Works of Tennyson
The Golden Year
L. 17

Thomson, Sir George

Science is essentially a search for truth.

The Inspiration of Science
Introduction (p. 1)

Toynbee, Arnold J.

The Truth apprehended by the Subconscious Psyche finds natural expression in Poetry; The Truth apprehended by the Intellect finds natural expression in Science . . .

In T. Dobzhansky
The Biology of Ultimate Concern
Chapter 6 (p. 115)

Unknown

The alternative to accepting that there is a strong measure of truth in science is to go back to blaming a witch when the cow is sick.

The Economist
The Nature of Knowledge (p. 103)
Volume 281, December 26, 1981–June 8, 1982

Vaihinger, H.

We have repeatedly insisted . . . that the boundary between truth and error is not a rigid one, and we were able ultimately to demonstrate that what we generally call truth, namely a conceptual world coinciding with the external world, *is merely the most expedient error*.

The Philosophy of 'As If'
Chapter XXIV (p. 108)

Weil, Simone

Truth is a radiant manifestation of reality.

The Need For Roots
Part Three (p. 253)

Wilkins, John

That the strangeness of this opinion is no sufficient reason why it should be rejected, because other certain truths have been formerly esteemed ridiculous, and great absurdities entertayned by common consent.

The Discovery of a World in the Moone
The First Proposition (p. 1)

UNDERSTANDING

Bacon, Francis

The eye of understanding is like the eye of the sense; for as you may see great objects through small crannies or levels, so you may see great axioms of nature through small and contemptible instances.

Sylva Sylvarum (p. 337)

Conrad, Joseph

Things and men have always a certain sense, a certain side by which they must be got hold of if one wants to obtain a solid grasp and a perfect command.

Under Western Eyes
4.1

Dahlberg, Edward

It takes a long time to understand nothing.

Reasons of the Heart
On Wisdom and Folly

Hazlitt, William

In what we really understand, we reason but little.

Literary Remains
On the Conduct of Life

Holmes, Oliver Wendell

A moment's insight is sometimes worth a life's experience.

The Professor at the Breakfast-Table
Iris, Her Book

Popper, Karl R.

The activity of understanding is, essentially, the same as that of all problem solving.

šitObjective Knowledge (p. 166)

BIBLIOGRAPHY

Adams, Henry. *The Education of Henry Adams*. The Modern Library, New York. 1946.

Agassiz, Luis. *Luis Agassiz: His Life and Correspondence*. Houghton, Mifflin and Company, Boston. 1886.

Altizer, Thomas J.J., Beardslee, William A. and Young, J. Harvey. *Truth, Myth, and Symbol*. Prentice-Hall, Inc., Englewood Cliffs. 1962.

Anderson, Maxwell. *Joan of Lorraine*. Anderson, Washington, D.C. 1947.

Appleyard, Bryan. *Understanding the Present*. Doubleday, New York. 1992.

Arber, Agnes. *The Mind and the Eye*. At the University Press, Cambridge. 1954.

Ardrey, Robert. *African Genesis*. Atheneum, New York. 1968.

Aristotle. 'Categories' in *Great Books of the Western World*. Volume 8. Encyclopaedia Britannica, Inc., Chicago. 1952.

Aristotle. 'Physics' in *Great Books of the Western World*. Volume 8. Encyclopaedia Britannica, Inc., Chicago. 1952.

Aristotle. 'Posterior Analytics' in *Great Books of the Western World*. Volume 8. Encyclopaedia Britannica, Inc., Chicago. 1952.

Arnheim, Rudolf. *Art and Visual Perception*. University of California Press, Berkeley. 1969.

Arnold, Thurman. *The Folklore of Capitalism*. Yale University Press, New Haven. 1937.

Aronowitz, Stanley. *Science as Power*. University of Minnesota Press, Minneapolis. 1988.

Arp, Halton. 'Letters' in *Science News*. Volume 140, Number 4. July 27, 1991.

Asimov, Isaac. *Of Time and Space and Other Things*. Avon Books, New York. 1965.

Attwell, Henry. *Thoughts From Ruskin*. Longmans, Green, and Co., New York. 1901.

Ayala, Francisco Jose and Dobzhansky, Theodosius. *Studies in the Philosophy of Biology*. Macmillan Press Limited, London. 1974.

Bacon, Francis. *Selected Writings of Francis Bacon*. Random House Inc., New York. 1955.

Bacon, Francis. 'Advancement in Learning' in *Great Books of the Western World*. Volume 30. Encyclopaedia Britannica, Inc., Chicago. 1952.

Bacon, Francis. 'Novum Organum' in *Great Books of the Western World*. Volume 30. Encyclopaedia Britannica, Inc., Chicago. 1952.

Balfour, Arthur James. *The Foundations of Belief*. Longmans, Green, and Co., London. 1912.

Barnett, Lincoln. 'The Meaning of Einstein's New Theory' in *Life*. January 9, 1950.

Barnett, Lincoln. 'J. Robert Oppenheimer' in *Life*. October 10, 1949.

Barnett, P.A. *Common Sense in Education and Teaching*. Longmans, Green, and Co., New York. 1899.

Barr, Amelia E. *All the Days of My Life*. D. Appleton and Company, New York. 1913.

Barrie, J.M. *The Admirable Crichton*. Hodder and Stoughton, London. 1961.

Barrow, John D. *The Artful Universe*. Clarendon Press, Oxford. 1995.

Barry, Frederick. *The Scientific Habit of Thought*. Columbia University Press, New York. 1927.

Bartlett, Elisha. *An Essay on the Philosophy of Medical Science*. Lea & Blanchard, Philadelphia. 1844.

Barzun, Jacques. *Science: The Glorious Entertainment*. Harper and Row, Publishers, New York. 1964.

Batchelor, G.K. 'Preoccupations of a Journal Editor' in *Journal of Fluid Mechanics*. Volume 106. 1981.

Bayliss, William Maddock. *Principles of General Physiology*. Longmans, Green, and Co., New York. 1920.

Bean, William. *Aphorisms from Latham*. The Prairie Press, Iowa City. 1962.

Beard, Charles A. 'Limitations to the Application of Social Science Implied in Recent Social Trends' in *Social Forces*. Volume XI, Number 4. May 1933.

Beattie, James. The Complete Poetical Works of Gray, Beattie, Black, Collins, Thomson, and Kirke White. James Blackwood & Co., London. No date.

Beaumont, William. *Experiment and Observations on the Gastric Juice and the Physiology of Digestion*. The C.V. Mosby Company, St. Louis. 1981.

Beck, Lewis White. 'The "Natural Science Ideal" in the Social Sciences' in *The Scientific Monthly*. Volume LXVIII. June 1949.

Bell, Eric T. *Debunking Science*. University of Washington Bookstore, Seattle. 1930.

Belloc, Hilaire. *Essays of a Catholic*. Books for Libraries Press, Inc., Freeport. 1967.

Bellone, Enrico. *A World on Paper*. The MIT Press, Cambridge. 1980.

Berger, Peter L. *A Rumor of Angels*. Doubleday & Company, Inc., Garden City. 1969.

Bernal, J.D. *Science in History*. Watts & Co., London. 1954.

Bernal, J.D. *The Social Function of Science*. The Macmillan Company, New York. 1939.

Bernard, Claude. *An Introduction to the Study of Experimental Medicine*. Henry Schuman, Inc. 1949.

Bernstein, Jeremy. *Experiencing Science*. Basic Books, Inc., Publishers, New York. 1978.

Besso, Michele. *Correspondance, 1903–1955*. Hermann, Paris. 1972.

Beveridge, W.I.B. *The Art of Scientific Investigation*. William Heinemann Ltd., Melbourne. 1950.

Billings, Josh. *Old Probabilities: Perhaps Rain—Perhaps Not*. Literature House, Upper Saddle River. 1970.

Blackie, John Stuart. *Musa Burschicosa*. Edmonston and Douglas, Edinburgh. 1869.

Blake, William. *Jerusalem*. Oxford University Press, London. 1913.

Blake, William. *The Complete Writings of William Blake*. Oxford University Press, London. 1966.

Blissett, Marlan. *Politics in Science*. Little, Brown and Company, Boston. 1972.

Bloch, Marc. *The Historian's Craft*. Alfred A. Knopf, New York. 1962.

Bloomfield, Leonard. 'Linguistic Aspects of Science' in *International Encyclopedia of Unified Science*. The University of Chicago Press, Chicago. 1939.

Bloor, David. *Knowledge and Social Imagery*. Routledge & Kegan Paul, London. 1976.

Bohm, David. 'On the Relationship between Methodology in Scientific Research and the Content of Scientific Knowledge' in *The British Journal for the Philosophy of Science*. Volume XII, Number 46. August 12, 1961.

Bohn, Henry G. *A Hand-Book of Proverbs*. George Bell & Sons, London. 1893.

Bohr, Niels. *Atomic Theory and the Description of Nature*. At the University Press, Cambridge. 1934.

Boltzmann, Ludwig. *Lectures on Gas Theory*. Translated by Stephen G. Brush. University of California Press, Berkeley. 1964.

Boltzmann, Ludwig. *Theoretical Physics and Philosophical Problems*. D. Reidel Publishing Company, Dordrecht. 1974.

Bondi, Hermann. *Relativity and Common Sense*. Anchor Books, Garden City. 1964.

Boole, George. *An Investigation of the Laws of Thought*. Dover Publications, Inc., New York. 1951.

Born, Max. *Natural Philosophy of Cause and Chance*. At the Clarendon Press, Oxford. 1949.

Born, Max. *Physics in My Generation*. Springer-Verlag, New York. 1969.

Boutroux, Émile. *Science & Religion in Contemporary Philosophy*. The Macmillan Company, New York. 1911.

Boycott, A.E. 'The Transition from Live to Dead' in *Nature*. Volume 123. January 19, 1929.

Boyd, T.A. *Professional Amateur, The Biography of Charles Franklin Kettering*. E.P. Dutton & Co., Inc., New York. 1957.

Bradley, A.C. *Oxford Lectures on Poetry*. Macmillan and Co., Limited, London. 1909.

Brain, Lord Walter Russell. 'Science and Antiscience' in *Science*. Volume 148, Number 3667. April 1965.

Braithwaite, Richard Bevan. *Scientific Explanation*. Harper & Brothers, New York. 1960.

Brewster, David. *Memoirs of the Life, Writings and Discoveries of Sir Isaac Newton*. Edmonston and Douglas, Edinburgh. 1860.

Bronowski, J. *A Sense of the Future*. The MIT Press, Cambridge. 1972.

Bronowski, J. *Science and Human Values*. Harper & Row, Publishers, New York. 1965.

Bronowski, J. *The Common Sense of Science*. William Heinemann Ltd., London. 1951.

Brooks, Paul. *The House of Life: Rachel Carson at Work*. Houghton Mifflin Company, Boston. 1972.

Brougham, Henry. 'The Bakerian Lecture on the Theory of Light and Colours' in *Edinburgh Review*. Volume 1. 1801–3.

Brouwer, L.E.J. 'Intuitionism and Formalism' in *Bulletin of the American Mathematical Society*. Volume 20. November 1913.

Brown, J. Howard. 'The Biological Approach to Bacteriology' in *Journal of Bacteriology*. Volume XXIII, Number 1. January 1932.

Brown, Kenneth A. *Inventors at Work*. Tempest Books of Microsoft Press, Redmond. 1988.

Browne, Sir Thomas. *Religio Medici*. At the University Press, Cambridge. 1955.

Brownell, Frederic. 'Heed That Hunch' in *The American Magazine*. December 1945.

Browning, Robert. *The Poems*. Yale University Press, New Haven. 1981.

Bruce, Philip Alexander. *History of the University of Virginia: 1819–1919*. Volume I. The Macmillan Company, New York. 1920.

Bube, Richard H. *The Encounter Between Christianity and Science*. William B. Eerdmans Publishing Company, Grand Rapids. 1968.

Buchanan, Scott. *Poetry and Mathematics*. The John Day Company, New York. 1929.

Buchler, Justus. *Philosophical Writings of Peirce*. Dover Publications, Inc., New York. No date.

Buck, Pearl S. *A Bridge for Passing*. The John Day Company, New York. 1962.

Buck, R. and Cohen, R. *Boston Studies in the Philosophy of Science*. D. Reidel Publishing Company, Dordrecht. 1971.

Buckham, John Wright. 'The Passing Scientific Era' in *The Century Illustrated Monthly Magazine*. August 1929.

Budworth, D. 'Science Policy Should Be About People' in *New Scientist*. Volume 69, Number 993. March 25, 1976.

Bulwer-Lytton, Edward. *Caxtoniana*. Harper & Brothers, Publishers, New York. 1864.

Bulwer-Lytton, Edward. *The Caxtons*. George Routledge and Sons, London. 1880.

Bunge, Mario. *Causality*. Harvard University Press, Cambridge. 1959.

Bunge, Mario. *The Myth of Simplicity*. Prentice-Hall, Inc., Englewood Cliffs. 1963.

Burhoe, R.W. 'The Source of Civilization in the Natural Selection of Coadapted Information in Genes and Culture' in *Zygon*. Volume 11, Number 3, 1976.

Burne, Glen S. *Selected Writings*. The University of Michigan Press, Ann Arbor. 1966.

Burroughs, John. 'In the Noon of Science' in *The Atlantic Monthly*. September 1912.

Burroughs, John. 'Scientific Faith' in *The Atlantic Monthly*. July 1915.

Burroughs, William. *The Adding Machine*. Seaver Books, New York. 1986.

Burton, Sir Richard. *The Kasidah of Hâjî Abdû El-Yezdî*. Thomas B. Mosher, Portland. 1898.

Burtt, Edwin Arthur. *The Metaphysical Foundations of Modern Physical Science*. The Humanities Press, Inc., New York. 1951.

Bush, Vannevar. *Endless Horizons*. Public Affairs Press, Washington, D.C. 1946.

Bush, Vannevar. *Modern Arms and Free Men*. Simon and Schuster, New York. 1949.

Bush, Vannevar. *Science Is Not Enough*. William Morrow & Company, Inc., New York. 1967.

Bushnell, Horace. 'Science and Religion' in *Putnam's Magazine*. Volume 1. 1868.

Butler, J.A.V. *Science and Human Life*. Basic Books, Inc., New York. 1957.

Butterfield, H. *The Origins of Modern Science 1300–1800*. The Macmillan Company, New York. 1961.

Byron, Lord George Gordon. *The Works of Lord Byron*. Oliver S. Felt, New York. No date.

Cabot, Richard C. *The Meaning of Right and Wrong*. The Macmillan Company, New York. 1933.

Calaprice, Alice. *The Quotable Einstein*. Princeton University Press, Princeton. 1996.

Calvin, William H. *The Cerebral Symphony*. Bantam Books, New York. 1990.

Campbell, Norman. *Physics: The Elements*. At the University Press, Cambridge. 1920.

Campbell, Norman. *What is Science?* Dover Publications, Inc., New York. 1952.

Campbell, Thomas. *Poetical Works*. Crosby, Nichols, Lee & Company, Boston. 1860.

Camus, Albert. *The Fall*. Vintage Books, New York. 1956.

Carlyle, Thomas. *Latter Day Pamphlets*. Scribner, Welford, and Company, New York. 1872.

Carlyle, Thomas. *The Nigger Question*. Apple-Century-Crofts, New York. 1971.

Casimir, Hendrik. *Haphazard Reality*. Harper & Row, Publishers, New York. 1983.

Cassirer, Ernst. *An Essay on Man: An Introduction to a Philosophy of Human Culture*. Yale University Press, New Haven. 1944.

Cervantes, Miguel de. 'Don Quixote del la Mancha' in *Great Books of the Western World*. Volume 29. Encyclopaedia Britannica, Inc., Chicago. 1952.

Chandler, Raymond. *The Notebooks of Raymond Chandler*. The Ecco Press, New York. 1976.

Chandrasekhar, S. *Truth and Beauty*. The University of Chicago Press, Chicago. 1987.

Chargaff, Erwin. 'Bitter Fruits from the Tree of Knowledge' in *Perspectives in Biology and Medicine*. Volume 16, Number 4. Summer 1975.

Chargaff, Erwin. 'Voices in the Labyrinth' in *Perspectives in Biology and Medicine*. Volume 18, Number 3. Spring 1975.

Chargaff, Erwin. 'Triviality in Science: A Brief Meditation on Fashions' in *Perspectives in Biology and Medicine*. Volume 19, Number 3. Spring 1976.

Chargaff, Erwin. 'In Praise of Smallness—How Can We Return to Small Science' in *Perspectives in Biology and Medicine*. Volume 23, Number 3. Spring 1980.

Chargaff, Erwin. 'Preface to a Grammar of Biology' in *Science*. Volume 172, Number 3984. May 1971.

Chargaff, Erwin. *Heraclitean Fire*. The Rockefeller University Press, New York. 1978.

Chernin, Kim. *The Obsession*. Harper & Row, Publishers, New York. 1981.

Chesterton, G.K. *All is Grist*. Methuen & Co., Ltd., London. 1933.

Chesterton, G.K. *All Things Considered*. John Lane Company, New York. 1916.

Chesterton, G.K. *Heretics*. John Lane Company, New York. 1909.

Chesterton, G.K. *Orthodoxy*. Garden City Publishing Company, Inc., Garden City. 1908.

Chesterton, G.K. *The Apostle and the Wild Ducks*. Paul Elek: London. 1975.
Chesterton, G.K. *The Defendant*. Dodd, Mead & Company, New York. 1904.
Chesterton, G.K. *The Everlasting Man*. Hodder & Stoughton Limited, London. 1925.
Chesterton, G.K. *The G.K. Chesterton Calendar*. Cecil Palmer & Hayward, London. 1916.
Chesterton, G.K. *The Resurrection of Rome*. Dodd, Mead & Company, New York. 1930.
Chesterton, G.K. *The Uses of Diversity*. Methuen & Co., Ltd., London. 1920.
Chesterton, G.K. *The Well and the Shallows*. Sheen and Ward, London. 1935.
Christy, Robert. *Proverbs, Maxims and Phrases*. Volume II. G.P. Putnam's Sons, New York. 1888.
Clarke, Arthur C. *Profiles of the Future*. Harper & Row, Publishers, New York. 1962.
Clifford, William Kingdon. *The Common Sense of the Exact Sciences*. Dover Publications, Inc., New York. 1955.
Cohen, H. Floris. *The Scientific Revolution*. The University of Chicago Press, Chicago. 1994.
Cohen, Morris Raphael. *Reason and Nature*. Free Press, Glencoe. 1953.
Cohen, Morris R. *The Meaning of Human History*. The Open Court Publishing Company, La Salle. 1947.
Cohen Morris R. and Nagel, Ernest. *An Introduction to Logic and Scientific Method*. Harcourt, Brace and Company, New York. 1934.
Coles, Abraham. *The Microcosm and Other Poems*. D. Appleton and Company, New York. 1881.
Collingwood, Robin George. *Speculum Mentis*. At the Clarendon Press, Oxford. 1924.
Collingwood, R.G. *The New Leviathan*. At the Clarendon Press, Oxford. 1942.
Collins, Wilkie. *Heart and Science*. Chatto & Windus, London. 1899.
Colodny, Robert G. *Mind and Cosmos*. University of Pittsburgh Press, Pittsburgh. 1966.
Comfort, Alex. 'On Physics and Biology: Getting Our Act Together' in *Perspectives in Biology and Medicine*. Volume 29, Number 1. Autumn 1985.
Committee on the Conduct of Science. *On Being a Scientist*. National Academy Press, Washington, D.C. 1989.
Commoner, Barry. *Science and Survival*. The Viking Press, New York. 1966.
Compton, Karl Taylor. *A Scientist Speaks*. Published by the Undergraduate Association, Massachusetts Institute of Technology, Cambridge. 1955.

Conant, James B. *Modern Science and Modern Man*. Columbia University Press, New York. 1952.

Conant, James B. *On Understanding Science*. Yale University Press, New Haven. 1947.

Conant, James B. *Science and Common Sense*. Yale University Press, New Haven. 1951.

Conference on Science, Philosophy and Religion. *Science, Philosophy and Religion*. Kraus Reprint Co., New York. 1971.

Congreve, William. *The Mourning Bride*. Dicks' Standard Plays. Number 99. No date.

Cornwell, John. *Nature's Imagination*. Oxford University Press, Oxford. 1995.

Coulson, C.A. *Science and Christian Belief*. The University of North Carolina Press, Chapel Hill. 1955.

Cowper, William. *Complete Poetical Works*. University Press, Oxford. 1913.

Crick, Francis. *Of Molecules and Men*. University of Washington Press, Seattle. 1966.

Crick, Francis. *What Mad Pursuit*. Basic Books, Inc., Publishers, New York. 1988.

Crombie, A.C. *Scientific Change*. Basic Books Inc., Publishers, New York. 1963.

Crosland, Maurice. *Gay-Lussac: Scientist and Bourgeois*. Cambridge University Press, Cambridge. 1978.

Crothers, Samuel McChord. *The Gentle Reader*. Books for Libraries Press, Freeport. 1972.

Cudmore, L.L. Larison. *The Center of Life*. The New York Times Book Co., New York. 1977.

Curie, Eve. *Madame Curie*. Doubleday, Doran & Company, Inc., Garden City. 1937.

Cushing, Harvey. *The Life of Sir William Osler*. At the Clarendon Press, Oxford. 1925.

Cushing, James T., Delaney, C.G. and Gutting, Gary M. *Science and Reality*. University of Notre Dame Press, Notre Dame. 1984.

Czarnomski, F.B. *The Wisdom of Winston Churchill*. George Allen and Unwin Ltd., London. 1956.

da Vinci, Leonardo. *Leonardo da Vinci's Notebooks*. Duckworth & Co., London. 1906.

da Vinci, Leonardo. *The Literary Works of Leonardo da Vinci*. Oxford University Press, London. 1939.

Daly, Reginald Aldworth. *Our Mobile Earth*. Charles Scribner's Sons, New York. 1926.

Darwin, Charles. 'The Descent of Man' in *Great Books of the Western World*. Volume 49. Encyclopaedia Britannica, Inc., Chicago. 1952.

Darwin, Charles. *More Letters of Charles Darwin*. Volume I. D. Appleton and Company, New York. 1903.

Darwin, Charles. *More Letters of Charles Darwin*. Volume II. D. Appleton and Company, New York. 1903.

Darwin, Francis. *The Life and Letters of Charles Darwin*. Volume II. D. Appleton and Company, New York. 1896.

Davie, Emily. *Profile of America*. Thomas Y. Crowell Company, New York. 1954.

Davies, P.C.W. and Brown, Julian. *Superstrings: A Theory of Everything?* Cambridge University Press, Cambridge. 1988.

Davis, B.D. 'Two Perspectives' in *Perspectives in Biology and Medicine*. Volume 35, Number 1. Autumn 1991.

Davis, Joel. *Alternate Realities*. Plenum Trade, New York. 1997.

Davis, Philip J. and Hersh, Reubens. *The Mathematical Experience*. Penguin Books, Ltd., Middlesex. 1983.

Davy, Humphrey. *Consolations in Travel*. Cassell & Company, Limited, London. 1889.

Davy, Humphrey. *Fragmentary Remains*. John Churchill, London. 1858.

Davy, Humphrey. *The Collected Works of Sir Humphrey Davy*. Volume VIII. Smith, Elder and Co., London. 1840.

Day Lewis, C. *Poetry For You*. Basil Blackwell, Oxford. 1945.

de Chardin, Pierre Teilhard. *The Phenomenon of Man*. Harper & Row, Publishers, New York. 1965.

de Jouvenel, Bertrand. *The Art of Conjecture*. Basic Books, New York. 1967.

de Kruif. 'America Comes through a Crisis' in *Saturday Evening Post*. May 13, 1933.

de Luc, Jean Andrew. *An Elementary Treatise on Geology*. Printed for F.C. and J. Rivington, London. 1809.

de Madariaga, Salvador. *Essays with a Purpose*. Hollis & Carter, London. 1954.

de Morgan, August. *A Budget of Paradoxes*. Volume I. The Open Court Publishing Co., Chicago. 1915.

de Reuck, Anthony, Goldsmith, Maurice and Knight, Julie. *Decision Making in National Science Policy*. Little, Brown and Company, Boston. 1968.

de Saint-Exupéry, Antoine. *The Little Prince*. Harcourt, Brace and Company, New York. 1943.

de Saint-Exupéry, Antoine. *Wind, Sand and Stars*. Harcourt, Brace Javanovich, Inc., New York. 1968.

Deason, Hilary J. *A Guide to Science Reading*. The New American Library, New York. 1964.

Delbrück, Max. 'A Physicist's Renewed Look at Biology: Twenty Years Later' in *Science*. Volume 168, Number 3937. 1970.

Delisle, Fanny. *A Study of Shelly's A Defence of Poetry*. Volume 1. University of Salzburg, Austria. 1974.

Deming, William Edward. *Statistical Adjustment of Data*. John Wiley & Sons, Inc., New York. 1943.

Descartes. 'Rules for the Direction of the Mind' in *Great Books of the Western World*. Volume 31. Encyclopaedia Britannica, Inc., Chicago. 1952.

Dewey, John. 'Common Sense and Science' in *The Journal of Philosophy*. Volume XLV, Number 8. April 18, 1948.

Dewey, John. *Common Faith*. Yale University Press, New Haven. 1934.

Dewey, John. *Freedom and Culture*. G.P. Putnam's Sons, New York. 1939.

Dewey, John. *Logic: The Theory of Inquiry*. Henry Holt & Co., New York. 1938.

Dewey, John. *Reconstruction in Philosophy*. Beacon Press, Boston. 1957.

Dewey, John. *The Quest for Certainty*. Minton, Balch, & Company, New York. 1929.

Dickens, Charles. *Pickwick Papers*. Chapman & Hall, London. 1836.

Dickinson, Emily. *The Complete Poems of Emily Dickinson*. Little, Brown and Company, Boston. 1960.

Dickinson, G. Lowes. *A Modern Symposium*. Doubleday, Page & Company, Garden City. 1920.

Dickinson, John P. *Science and Scientific Researchers in Modern Society*. UNESCO, Paris. 1984.

Dingle, Herbert. *The Scientific Adventure*. Sir Isaac Pitman & Sons, Ltd., London. 1952.

Disraeli, Benjamin. *Coningsby*. Penguin Books, New York. 1982.

Disraeli, Benjamin. *Lothair*. Longmans, Green, and Co., London. 1970.

Dobie, J. Frank. *The Voice of the Coyote*. University of Nebraska Press. 1962.

Dobzhansky, Theodosius. *The Biology of Ultimate Concern*. The New American Library, New York. 1967.

Dornan, Christopher. 'Some Problems of Conceptualizing the Issues of "Science and the Media" in *Critical Studies in Mass Communication*. Volume 7, Number 1. March 1990.

Douglas, A. Vibert. 'From Atoms to Stars' in *The Atlantic Monthly*. August 1929.

Dovring, Folke. *Knowledge and Ignorance*. Praeger, Westport. 1998.

Doyle, Sir Arthur Conan. *The Complete Sherlock Holmes*. Doubleday & Company, Inc., Garden City. 1960.

Drake, Daniel. *Introductory Lectures, on the Means of Promoting the Intellectual Improvement of Students*. 1844.

Dretske, Fred I. *Knowledge & the Flow of Information*. The MIT Press, Cambridge. 1981.

Drexler, K. Eric. *Engines of Creation*. Anchor Press/Doubleday, Garden City. 1986.

Drummond, Sir William. *Academical Questions*. Volume I. W. Bulmer and Co., London. 1805.

du Noüy, Lecomte. *Between Knowing and Believing*. Translated by Mary Lecomte du Noüy. David McKay Company, Inc., New York. 1966.

du Noüy, Lecomte. *Human Destiny*. Longmans, Green and Co., New York. 1947.

du Noüy, Lecomte. *The Road to Reason*. Longmans, Green and Company, New York. 1949.

Dubos, René. 'Two Perspectives' in *Perspectives in Biology and Medicine*. Volume 35, Number 1. Autumn 1991.

Dubos, René. *Louis Pasteur: Free Lance of Science*. Charles Scribner's Sons, New York. 1976.

Dubos, René. *Pasteur and Modern Science*. Science Tech Publishers, Madison. 1988.

Duhem, Pierre. *The Aim and Structure of Physical Theory*. Princeton University Press, Princeton. 1954.

Dukas, Helen and Hoffmann, Banesh. *Albert Einstein: The Human Side*. Princeton University Press, Princeton. 1979.

Dumas, Maurice. *Scientific Instruments of the 17th and 18th Centuries and Their Makers*. B.T. Batsford, London. 1972.

Dunlap, Knight. 'The Outlook for Psychology' in *Science*. Volume LXIX, Number 1782. February 22, 1929.

Dunne, Finley Peter. *Mr Dooley On Making a Will*. Charles Scribner's Sons, New York. 1919.

Durant, Will. *The Story of Philosophy*. Garden City Publishing Co., Inc., Garden City. 1927.

Duschl, R. *Restructuring Science and Education*. Teachers College Press, New York. 1990.

Dyson, Freeman. *Disturbing the Universe*. Harper & Row, Publishers, New York. 1979.

Dyson, Freeman. *Infinite in All Directions*. Harper and Row, Publishers, New York. 1988.

Eddington, Arthur Stanley. *Science and the Unseen World*. The Macmillan Company, New York. 1929.

Eddington, Arthur Stanley. *The Nature of the Physical World*. The Macmillan Company, New York. 1948.

Eddy, Mary Baker. *Science and Health with Key to the Scriptures*. Christian Science Publishing Society, Boston. 1918.

Edelman, Gerald. *Bright Air, Brilliant Fire*. Basic Books, New York. 1992.

Efron, Bradley, and Tibshirani, Robert J. *An Introduction to the Bootstrap*. Chapman & Hall, New York. 1993.

Egler, Frank E. *The Way of Science*. Hafner Publishing Company, New York. 1970.

Einstein, Albert. 'Science and God: A Dialog' in *Forum*. Volume 83. June 1930.

Einstein, Albert. 'Considerations Concerning the Fundaments of Theoretical Physics' in *Science*. Volume 91, Number 2369. May 24, 1940.

Einstein, Albert. 'Einstein Seeks Lack in Applying Science' in *The New York Times*. February 17, 1931.

Einstein, Albert. *Cosmic Religion*. Covici Friede, Publishers, New York. 1931.

Einstein, Albert. *Ideas and Opinions*. Bonanza Books, New York. 1954.

Einstein, Albert. *Out of My Later Years*. Philosophical Library, New York. 1950.

Einstein, Albert. *The World As I See It*. Covici, Friede Publishers, New York. 1934.

Einstein, Albert and Infeld, Leopold. *The Evolution of Physics*. Simon and Schuster, New York. 1938.

Eiseley, Loren. *The Star Thrower*. Times Books, New York. 1978.

Eisenberg, Anne. 'The Unabomber and the Bland Decade' in *Scientific American*. Volume 274, Number 4. April 1998.

Eisenschiml, Otto. *The Art of Worldly Wisdom*. Duell, Sloan and Pearce, New York. 1947.

Eliot, George. *Middlemarch*. Oxford University Press, Oxford. 1988.

Eliot, T.S. *Collected Poems 1909–1935*. Harcourt, Brace and Company, New York. 1936.

Emerson, Ralph Waldo. *The Journals and Miscellaneous Notebooks of Ralph Waldo Emerson*. Volume III. The Belknap Press of Harvard University Press, Cambridge. 1963.

Epps, John. *The Life of John Walker, M.D.* Whittaker, Treacher, and Co., London. 1831.

Esquivel, Laura. *Like Water for Chocolate*. Translated by Carol Christensen and Thomas Christensen. Doubleday, New York. 1992.

Faber, Harold. *The Book of Laws*. Times Books. 1979.

Fabings, Harold and Marr, Ray. *Fischerisms*. C.C. Thomas, Springfield. 1937.

Farb, Peter. *Man's Rise to Civilization*. E.P. Dutton & Co., Inc., New York. 1968.

Feibleman, James K. 'Pure Science, Applied Science, Technology, Engineering: An Attempt at Definitions' in *Technology and Culture*. Volume II, Number 4. Fall 1961.

Feigl, H. 'Naturalism *and* Humanism' in *American Quarterly*. Volume I, Number 2. Summer 1949.

Ferré, Nels F.S. *Faith and Reason*. Harper & Brothers Publishers, New York. 1946.

Feyerabend, Paul. 'Realism and the Historicity of Knowledge' in *The Journal of Philosophy*. Volume LXXXVI, Number 8. 1989.

Feyerabend, Paul. *Against Method*. NLB, London. 1975.

Feynman, Richard P. *What Do You Care What Other People Think?* W.W. Norton & Company, New York. 1988.

Finch, James Kip. 'Engineering and Science' in *Technology and Culture*. Fall 1961.

Finch, James Kip. *Engineering and Western Civilization*. McGraw-Hill Book Company, Inc., New York. 1951.

Fischer, David Hackett. *Historian's Fallacies*. Harper & Row, Publishers, New York. 1970.

Fiske, John. *The Destiny of Man*. Houghton, Mifflin and Company, Boston. 1884.

Fitzgerald, Penelope. *The Gate of Angels*. Doubleday, New York. 1990.

Flammarion, Camille. *Popular Astronomy*. Chatto & Windus, London. 1894.

Flaubert, Gustave. *Dictionary of Accepted Ideas*. Max Reinhardt, London. 1954.

Flexner, Abraham. *Medical Education*. The Macmillan Company, New York. 1925.

Flexner, Abraham. *Universities*. Oxford University Press, London. 1968.

Foerster, Norman. *Humanism and America*. Farrar and Rinehart, Incorporated, New York. 1930.

Fohling, D. and Oparin, A. *Molecular Evolution: Prebiological and Biological*. Plenum Press, New York. 1972.

Forde, Victoria. *The Poetry of Basil Bunting*. Bloodaxe Books, Newcastle. 1991.

Fosdick, Harry Emerson. *Adventurous Religion*. Harper & Brothers, New York. 1926.

Fothergill, Philip G. *Life and Its Origin*. Sheed and Ward, London. 1958.

Fox, Russell, Garbuny, Max and Hooke, Robert. *The Science of Science*. Walker and Company, New York. 1963.

Fox, Sidney W. *The Origins of Prebiological Systems*. Academic Press, New York. 1965.

France, Anatole. *Penguin Island*. The Heritage Press, New York. 1947.

France, Anatole. *The Crime of Sylvestre Bonnard*. Translated by Lafcadio Hearn. Harper & Brothers, New York. 1890.

France, Anatole. 'The Opinions of Jérôme Coignard' in *The Authorized English Translation of the Novels and Short Stories of Anatole France*. Volume II. Parke, Austin and Lipscomb Inc., New York. No date.

Frank, Philipp. *Relativity—A Richer Truth*. Jonathan Cape, London. 1951.

Frazier, A.W. 'The Practical Side of Creativity' in *Hydrocarbon Processing*. Volume 45, Number 1. January 1966.

Fredrickson, A.G. 'The Dilemma of Innovating Societies' in *Chemical Engineering Education*. Summer 1969.

Free, E.E. 'The Electrical Brains in the Telephone' in *The World's Work*. Volume LIII, Number 4. February 1927.

Freud, Sigmund. *The Future of an Illusion*. Doubleday & Company, Inc., Garden City. 1961.

Friedenberg, Edgar Z. *The Vanishing Adolescent*. Beacon Press, Beacon Hill. 1959.

Friedman, Milton. *Essays in Positive Economics*. The University of Chicago Press, Chicago. 1953.

Friend, Julius W. and Feibleman, James. *What Science Really Means*. George Allen & Unwin Ltd., London. 1937.

Fromm, Erich. *The Sane Society*. Fawcett Publications, Inc., Greenwich. 1955.

Fuller, R. Buckminster. *Nine Chains to the Moon*. Southern Illinois University Press, Carbondale. 1963.

Fuller, Thomas. *The Holy State and the Profane State*. Printed for Thomas Tegg, London. 1841.

Galbraith, John Kenneth. *The New Industrial State*. The New American Library, New York. 1967.

Galsworthy, John. 'Eldest Son' in *Plays*. Second Edition. Charles Scribner's Sons, New York. 1921.

Galton, Francis. 'Scientific Achievement and Aptitude' in *Nature*. Volume 118, Number 2976. November 13, 1926.

Gardner, Martin. *Fads and Fallacies*. Dover Publications, New York. 1957.

Garman, Eliza Miner. *Letters, Lectures and Addresses of Charles Edward Garman*. Houghton Mifflin Company, Boston. 1909.

George, William H. *The Scientist in Action*. William & Norgate Ltd., London. 1936.

Gerth, H.H. and Mills, C. Wright. *From Max Weber*. Oxford University Press, New York. 1946.

Giddings, Franklin H. 'Societal Variables' in *The Journal of Social Forces*. Volume I, Number 4. May 1923.

Giere, Ronald N. *Explaining Science: A Cognitive Approach*. University of Chicago Press, Chicago. 1988.

Gilkey, Langdon. *Creationism on Trial: Evolution and God at Little Rock*. Winston Press, Minneapolis. 1985.

Gleick, J. *Chaos: Making a New Science*. Viking Penguin Inc., New York. 1987.

Godlovitch, Stanley and Rosiland, and Harris, John. *Animals, Men and Morals*. Taplinger Publishing Company, New York. 1972.

Goethe, Johann Wolfgang von. *Botanical Writings*. Translated by Bertha Mueller. University of Hawaii Press, Honolulu. 1952.

Goethe, Johann Wolfgang von. *Scientific Studies*. Volume 12. Suhrkamp Publishers, New York. 1988.

Goldenweiser, Alexander. *Robots or Gods*. Alfred A. Knopf, New York. 1931.

Goldsmith, Maurice and Mackay, Alan. *Society and Science*. Simon and Schuster, New York. No date.

Goodspeed, Edgar J. *The Four Pillars of Democracy*. Harper & Brothers Publishers, New York. 1940.

Goran, Morris. *Science and Anti-Science*. Ann Arbor Science Publishers Inc., Ann Arbor. 1974.

Gornick, Vivian. *Women in Science*. Simon and Schuster, New York. 1983.

Gould, Laurence M. 'Science and the Culture of Our Times' in *UNESCO Courier*. February 1968.

Gould, Stephen Jay. *The Mismeasure of Man*. W.W. Norton & Company, New York. 1981.

Graham, Loren R. *Between Science and Values*. Columbia University Press, New York. 1981.

Graham, Loren R. *The Soviet Academy of Sciences and the Communist Party, 1927–1932*. Princeton University Press, Princeton. 1967.

Gray, George. 'The Riddle of Our Reddening Skies' in *Harper's Monthly Magazine*. July 1937.

Gray, Thomas. *The Complete Poetical Works of Gray, Beattie, Blair, Collins, Thomson, and Kirke White*. James Blackwood & Co., London. No date.

Green, Celia. *The Decline and Fall of Science*. Hamish Hamilton, London. 1976.

Gregg, Alan. *The Furtherance of Medical Research*. Yale University Press, New Haven. 1941.

Grinnel, Frederick. 'Complementarity: An Approach to Understanding the Relationship between Science and Religion' in *Perspectives in Biology and Medicine*. Volume 29, Number 2. Winter 1986.

Groen, Janny, Smit, Eefke and Juurd Eijsvoogel. *The Discipline of Curiosity*. Elsvier Science Publishers, Amsterdam. 1990.

Grossman, Ron. 'Strong Words' in *Chicago Tribune*. January 1, 1993.

Gruenberg, Benjamin. *Science and the Public Mind*. McGraw-Hill Book Co., Inc., New York. 1935.

Hacking, Ian. *The Emergence of Probability*. Cambridge University Press, Cambridge. 1978.

Hager, Thomas. *Force of Nature: the Life of Linus Pauling*. Simon & Schuster, New York. 1995.

Hall, A. R. 'Can the History of Science be History?' in *British Journal of the History of Science*. Volume IV, Part III, Number 15. June 1969.

Hall, A.R. *The Scientific Revolution 1500–1800*. Longmans, Green and Co., London. 1954.

Hall, David. 'Letters' in *New Scientist*. Volume 142, Number 1922. April 23, 1994.

Hamilton, Sir William Rowan. 'The Imagination in Mathematics' in *North American Review*. Volume 85, Number 176. July 1857.

Harding, Rosamond E. *An Anatomy of Inspiration*. W. Heffer & Sons Ltd., Cambridge. 1940.

Hardy, G.H. *A Mathematician's Apology*. At the University Press, Cambridge. 1967.

Hardy, Thomas. *Collected Letters*. Volume 3. Clarendon Press, Oxford. 1982.

Harré, Rom. *Problems of Scientific Revolution*. Clarendon Press, Oxford. 1975.

Harré, Rom. *Varieties of Realism*. Basil Blackwell, Oxford. 1986.

Harrington, John. *Dance of the Continents*. J.P. Tarcher, Inc., Los Angeles. 1983.

Harris, Errole E. *Hypothesis and Perception*. George Allen & Unwin Ltd., London. 1970.

Harris, Ralph. *Growth, Advertising, and the Consumer*. The Institute of Economic Affairs Ltd. 1964.

Harrison, Jane Ellen. *Ancient Art and Ritual*. Thornton Butterworth Ltd., London. 1935.

Harth, Erich. *The Creative Loop: How the Brain Makes a Mind*. Addison-Wesley Publishing Company, Reading. 1993.

Harvey, William. 'An Anatomical Disquisition on the Motion of the Heart and Blood in Animals' in *Great Books of the Western World*. Volume 28. Encyclopaedia Britannica, Inc., Chicago. 1952.

Harvey, William. 'Anatomical Exercises on the Generation of Animals' in *Great Books of the Western World*. Volume 28. Encyclopaedia Britannica, Inc., Chicago. 1952.

Harwit, Martin. *Cosmic Discovery*. Basic Books, Inc., Publishers, New York. 1981.

Hawking, Stephen. *A Brief History of Time*. Bantam Books, New York. 1988.

Hazlitt, William. *Characteristics*. George Bell & Sons, London. 1884.

Heaviside, Oliver. *Electromagnetic Theory*. Volume III. Benn Brothers, Limited, London. 1922.

Heinlein, Robert A. *The Cat Who Walks Through Walls*. Putnam, New York. 1985.

Heisenberg, Werner. *Across the Frontiers*. Harper & Row, Publishers, New York. 1974.

Heisenberg, Werner. *Physics and Beyond*. Harper & Row, Publishers. New York. 1971.

Heisenberg, Werner. *Physics and Philosophy*. Harper & Brothers Publishers, New York. 1958.

Henderson, Lawrence J. *The Fitness of the Environment*. The Macmillan Company, New York. 1913.

Herold, J. Christopher. *The Mind of Napoleon*. Columbia University Press, New York. 1955.

Herschel, J.F.W. *A Preliminary Discourse on The Study of Natural Philosophy*. Longman, Rees, Orme, Brown, and Green, London. 1831.

Hertz, Richard C. *The American Jew in Search of Himself*. Bloch Publishing Company, New York. 1962.

Herzen, Alexander. *My Past and Thoughts*. Translated by Constance Garnett. Alfred A. Knopf, New York. 1973.

Herzen, Alexander. *Selected Philosophical Works*. Foreign Languages Publishing House, Moscow. 1956.

Hesse, Mary B. 'Operational Definition and Analogy in Physical Theories' in *British Journal for the Philosophy of Science*. Volume II, Number 8. February 1952.

Hesse, Mary B. *Revolutions and Reconstructions in the Philosophy of Science*. Indiana University Press, Bloomington. 1980.

Heywood, Robert B. *The Works of the Mind*. The University of Chicago Press, Chicago. 1947.

Highet, Gilbert. *Man's Unconquerable Mind*. Columbia University Press, New York. 1954.

Hilbert, David. 'Hilbert: Mathematical Problems' in *Bulletin of the American Mathematical Society*. Volume 8, 2nd series. October 1901–July 1902.

Hill, D.W. *Science*. Chemical Publishing Co., Inc., Brooklyn, 1946.

Hilts, Philip J. *Scientific Temperaments: Three Lives in Contemporary Science*. Simon and Schuster, New York. 1982.

Hinshelwood, C.N. *The Structure of Physical Chemistry*. At the Clarendon Press, Oxford. 1951.

Hoagland, Hudson. 'Science and the New Humanism' in *Science*. Volume 143. 1964.

Hobbes, Thomas. 'Leviathan' in *Great Books of the Western World*. Volume 23. Encyclopaedia Britannica, Inc., Chicago. 1952.

Hoefler, Don C. 'But You Don't Understand the Problem' in *Electronic News*. July 17, 1967.

Hoffer, Eric. *The Passionate State of Mind*. Harper & Brothers, New York. 1955.

Hoffmann, Banesh. *Albert Einstein: Creator and Rebel*. A Plume Book, New York. 1972.

Hogan, James P. *Code of the Life Maker*. Ballantine Books, New York. 1983.

Holmes, Oliver W. *The Autocrat of the Breakfast-Table*. Houghton Mifflin, Boston. 1894.

Holmes, Oliver W. *The Complete Poetical Works of Oliver Wendell Holmes*. The Riverside Press, Cambridge. 1899.

Holmes, Oliver W. *The Poet at the Breakfast-Table*. Houghton, Mifflin and Company, Boston. 1884.

Holmes, Oliver W. *The Writings of Oliver Wendell Holmes*. The Riverside Press, Cambridge. 1911.

Holstrom, J. Edwin. *Records and Research in Engineering and Industrial Science*. Chapman & Hall Ltd., London. 1956.

Holton, Gerald. *Thematic Origins of Scientific Thought*. Harvard University Press, Cambridge. 1973.

Holton, Gerald and Roller, Duane H.D. *Foundations of Modern Physical Science*. Addison-Wesley Publishing Company, Inc., Reading. 1958.

Hooykaas, R. *Religion and the Rise of Modern Science*. Scottish Academic Press, Edinburgh. 1972.

Horgan, John. 'Quantum Consciousness' in *Scientific American*. Volume 261, Number 5. November 1989.

Horgan, John. *The End of Science*. Addison-Wesley Publishing Company, Inc., Reading. 1996.

Hoyle, Fred. *Of Men and Galaxies*. University of Washington Press, Seattle. 1964.

Hsu, Francis L.K. *Health, Culture and Community*. Russell Sage Foundation, New York. 1955.

Hubble, Edwin. *The Nature of Science*. The Huntington Library, San Marino. 1954.

Hubble, Ruth, Henifin, Mary Sue and Fried, Barbara. *Women Look at Biology Looking at Women*. Schenkman Publishing Co., Cambridge. 1979.

Huizinga, John. *Homo Ludens*. Beacon Press, Boston. 1950.

Hume, David. 'Concerning Human Understanding' in *Great Books of the Western World*. Volume 35. Encyclopaedia Britannica, Inc., Chicago. 1952.

Husserl, Edmund. *The Crisis of European Sciences and Transcendental Phenomenology*. Northwestern University Press, Evanston. 1970.

Hutchins, Robert M. and Adler, Mortimer J. *The Great Ideas Today 1974*. Encyclopaedia Britannica, Inc., Chicago. 1974.

Huxley, Aldous. *Along the Road*. Chatto & Windus, London. 1925.

Huxley, Aldous. *Ends and Means*. Harper & Brothers Publishers, New York. 1937.

Huxley, Aldous. *Literature and Science*. Harper & Row, Publishers, New York. 1963.

Huxley, Aldous. *Science, Liberty and Peace*. Harper & Brothers Publishers, New York. 1946.

Huxley, Aldous. *Texts and Pretexts*. Chatto & Windus, London. 1959.

Huxley, Aldous. *Tomorrow and Tomorrow and Tomorrow*. Harper and Row, Publishers, New York. 1972.

Huxley, Julian. 'Will Science Destroy Religion?' in *Harper's Monthly Magazine*. April 1926.

Huxley, Julian. 'Searching for the Elixir of Life' in *The Century Illustrated Monthly Magazine*. Volume 103, Number 4. February 1922.

Huxley, Julian. *What Dare I Think?* Chatto & Windus, London. 1932.

Huxley, Leonard. *Life and Letters of Thomas Henry Huxley*. D. Appleton and Company, New York. 1901.

Huxley, Thomas. *Collected Essays*. Volumes I–VIII. Greenwood Press, New York. 1968.

Huxley, Thomas. *Science and Education*. D. Appleton, New York. 1894.

Huxley, Thomas H. and Huxley, Julian. *Evolution and Ethics 1893–1943*. The Pilot Press, Ltd., London. 1947.

Infeld, Leopold. *Quest—An Autobiography*. Chelsea Publishing Company, New York. 1980.

Inge, William Ralph. *Outspoken Essays*. Second series. Longmans, Green and Co., New York. 1922.

Ingersoll, Robert G. *On the Gods and Other Essays*. Prometheus Books, Buffalo. 1990.

Ingle, Dwight J. *Principles of Research in Biology and Medicine*. J.B. Lippincott Company, Philadelphia. 1958.

Inose, Hiroshi and Pierce, John R. *Information, Technology and Civilization*. W.H. Freeman and Company, New York. 1984.

Jacks, L.P. 'Is There a Foolproof Science?' In *The Atlantic Monthly*. February 1924.

Jacob, François. *The Possible and the Actual*. Pantheon Books, New York. 1982.

James, William. *The Principles of Psychology*. Volume II. Henry Holt and Company, New York. 1890.

James, William. *The Varieties of Religious Experience*. The Modern Library, New York. 1929.

James, William. *The Will to Believe and Other Essays in Popular Philosophy*. Longmans Green and Co., New York. 1896.

Jastrow, Robert. *God and the Astronomers*. Warner Books, New York. 1980.

Jeans, Sir James. 'The Mathematical Aspect of the Universe' in *Philosophy*. Volume VII, Number 25. January 1932.

Jeans, Sir James. *The Mysterious Universe*. The Macmillan Company, New York. 1932.

Jeans, Sir James. *Physics and Philosophy*. Dover Publications, Inc., New York. 1981.

Jeffreys, H. *Theory of Probability*. At the Clarendon Press, Oxford. 1961.

Jenson, Elwood V. 'The Science of Science' in *Perspectives in Biology and Medicine*. Volume 12, Number 2. Winter 1969.

Jevons, William Stanley. *The Principles of Science*. Dover Publications, Inc., New York. 1958.

Joad, C.E.M. *Guide to Modern Thought*. Frederick A. Stokes Company, New York. 1933.

Johnson, George. *Fire in the Mind*. Alfred A. Knopf, New York. 1995.

Johnson, George. *In the Palaces of Memory*. Alfred A. Knopf, New York. 1991.

Jonson, Ben. *Discoveries Made Upon Man and Matter*. Cassell & Company Ltd., London. 1889.

Joubert, Joseph. *Pensées*. Books for Libraries, Freeport. 1972.

Judson, Horace Freeland. *The Eighth Day of Creation*. Simon and Schuster, New York. 1979.

Jungk, Robert. *Brighter than a Thousand Suns*. Harcourt, Brace and Company, New York. 1956.

Kant, Immanuel. 'The Critique of Pure Reason' in *Great Books of the Western World*. Volume 42. Encyclopaedia Britannica, Inc., Chicago. 1952.

Kant, Immanuel. *Universal Natural History and the Theory of the Heavens*. The University of Michigan Press, Ann Arbor. 1969.

Kapitza, P.L. 'Address to the Royal Society in Honour of Lord Rutherford' in *Nature*. Volume 210. May 17, 1966.

Kaplan, Abraham. *The Conduct of Inquiry*. Chandler Publishing Co., San Francisco. 1964.

Karanikas, Alexander. *Tillers of a Myth*. The University of Wisconsin Press, Madison. 1966.

Kass-Simon, G. and Farnes, Patricia. *Women of Science*. Indiana University Press, Bloomington. 1990.

Katscher, F. 'Correspondence' in *Nature*. Volume 363. June 3, 1993.

Keats, John. *Complete Poems*. Harvard University Press, Cambridge. 1982.

Kepes, Gyorgy. *The New Landscape in Art and Science*. Paul Theobald and Co., Chicago. 1956.

Kettering, Charles F. and Smith, Beverly. 'Ten Paths to Fame and Fortune' in *The American Magazine*. December 1937.

Keynes, Geoffrey and Hill, Brian. *Samuel Butler's Notebooks*. Jonathan Cape, London. 1951.

Keyser, Cassius J. *Mathematics*. The Columbia University Press, New York. 1907.

Keyser, Cassius J. *Mole Philosophy & Other Essays*. E.P. Dutton & Company, New York. 1927.

King, Martin Luther, Jr. *Strength To Love*. Harper & Row, Publishers, New York. 1963.

Kingsley, Charles. *Health and Education*. W. Isbister & Co., London. 1874.

Kipling, Rudyard. *Rudyard Kipling's Verse: Inclusive Edition*. Hodder and Stoughton Ltd., London. 1912.

Kirkpatrick, Clifford. *Religion in Human Affairs*. John Wiley & Sons, Inc., New York. 1929.

Kistiakowski, George B. 'Science and Foreign Affairs' in *Bulletin of the Atomic Scientists*. Volume 16, Number 4. April 1960.

Klee, Paul. *The Diaries of Paul Klee 1898–1918*. Diary III. University of California Press, Berkeley. 1964.

Kline, Morris. *Mathematics and the Search for Knowledge*. Oxford University Press, New York. 1985.

Kline, Morris. *Mathematics in Western Culture*. Oxford University Press, New York. 1953.

Koestler, Arthur. *Insight and Outlook*. Macmillan & Co. Ltd., London. 1949.

Koestler, Arthur. *The Roots of Coincidence*. Vintage Books, New York. 1972.

Koestler, Arthur and Smythies, J.R. *Beyond Reductionism*. The Macmillan Company, New York. 1970.

Köhler, Wolfgang. *Dynamics in Psychology*. Liveright Publishing Corporation, New York. 1940.

Kough, A. 'The Progress of Physiology' in *Science*. Volume 70, Number 1808. August 1929.

Kragh, Helge. *Dirac: A Scientific Biography*. Cambridge University Press, Cambridge. 1990.

Kroeber, A.L. *The Nature of Culture*. The University of Chicago Press, Chicago. 1952.

Kronenberger, Louis. *Company Manners*. The Bobbs-Merrill Company, Inc., Indianapolis. 1954.

Krutch, Joseph Wood. *Human Nature and the Human Condition*. Random House, New York. 1959.

Krutch, Joseph Wood. *The Measure of Man*. The Bobbs-Merrill Company, Inc., Indianapolis. 1954.

Krutch, Joseph Wood. *The Modern Temper*. Harcourt, Brace and Company, New York. 1929.

Kuhn, Thomas S. *International Encyclopedia of the Social Sciences*. Volume 14. The Macmillan Company and the Green Press. 1968.

Kuhn, Thomas S. *The Structure of Scientific Revolutions*. Second Edition. The University of Chicago Press, Chicago. 1970.

Lakatos, I. 'Criticism and the Methodology of Scientific Research Programmes' in *Proceedings of the Aristotelian Society*. Volume 69. No date.

Lakatos, Imre and Musgrave, Alan. *Criticism and the Growth of Knowledge*. Cambridge University Press, London. 1970.

Lamb, Charles. *The Essays of Elia*. J.M. Dent & Sons Limited, London. 1939.

Landheer, Barth. 'Presupposition in the Social Sciences' in *American Journal of Sociology*. Volume XXXVII. January 1932.

Lane, Albert. *Elbert Hubbard and His Work*. The Blanchard Press, Worcester. 1901.

Langer, Susanne K. *Philosophy in a New Key*. Harvard University Press, Cambridge. 1957.

Lapp, Ralph E. *The New Priesthood*. Harper & Row, Publishers, New York. 1965.

Larrabee, Eric. 'Science and the Common Reader' in *Commentary*. June 1966.

Larrabee, Harold A. *Reliable Knowledge*. Houghton Mifflin Company, Boston. 1945.

Latour, Bruno. *Science in Action*. Harvard University Press, Cambridge. 1987.

Latour, Bruno and Woolgar, Steve. *Laboratory Life: The Social Construction of Scientific Facts*. Sage Publications, Beverly Hills. 1979.

Laudan, Larry. *Progress and Its Problems*. University of California Press, Berkeley. 1977.

Le Guin, Ursula K. *The Language of the Night*. G.P. Putnam's Sons, New York. 1979.

Lebowitz, Fran. *Metropolitan Life*. Fawcett Columbine, New York. 1978.

Lee, Philip R., Ornstein, Robert E., Galin, David, Deikman, Arthur and Tart, Charles T. *Symposium on Consciousness*. The Viking Press, New York. 1976.

Lehrer, Keith. *Knowledge*. Clarendon Press, Oxford. 1974.

Lehrs, Ernst. *Man or Matter*. Faber and Faber Ltd., London. 1951.

Leonard, Jonathan Norton. 'Steinmetz, Jove of Science, Part II' in *The World's Work*. February 1929.

Lerner, Max. *Actions and Passions*. Simon and Schuster, New York. 1949.

Levin, Max. 'Our Debt to Hughlings Jackson' in *Journal of the American Medical Association*. Volume 191, Number 12. March 22, 1965.

Levine, George. *One Culture: Essays in Science and Literature*. The University of Wisconsin Press, Madison. 1987.

Lewin, Roger. *Complexity*. Macmillan Publishing Company, New York. 1992.

Lewis, C.S. *The Pilgrim's Regress: An Allegorical Apology for Christianity, Reason and Romanticism*. Sheed & Ward Inc., New York. 1935.

Lewis, Gilbert N. *The Anatomy of Science*. Yale University Press, New Haven. 1926.

Lewis, Sinclair. *Arrowsmith*. The Modern Library, New York. 1925.

Lewis, Wyndham. *The Art of Being Ruled*. Chatto and Windus, London. 1926.

Lévi-Strauss, Claude. *The Raw and the Cooked*. Harper & Row, Publishers, New York. 1969.

Levy, Hyman. *The University of Science*. The Century Co., New York. 1933.

Lichtenberg, Georg Christoph. *Lichtenberg: Aphorisms & Letters*. Jonathan Cape, London. 1969.

Lightman, Alan and Brawer, Roberta. *Origins*. Harvard University Press, Cambridge. 1990.

Loehle, Craig. 'A Guide to Increased Creativity in Research—Inspiration or Perspiration?' in *BioScience*. Volume 40, Number 2. February 1990.

Lorand, Arnold. *Old Age Deferred*. F.A. Davis Company, Publishers, Philadelphia. 1911.

Lorenz, Konrad. *On Aggression*. Harcourt, Brace & World, Inc., New York. 1963.

Lovecraft, H.P. and de Castro, Adolphe. *The Horror in the Museum and Other Revisions*. Arkham House: Publishers, Sauk City. 1970.

Lubbock, John. *The Beauties of Nature*. Macmillan and Co., Limited, London. 1900.

Lundberg, G.A. *Can Science Save Us?* Longmans, Green and Co., New York. 1947.

Luria, S.E. *A Slot Machine, A Broken Test Tube*. Harper & Row, Publishers, New York. 1984.

MacCurdy, Edward. *The Notebooks of Leonardo da Vinci*. Volume 1. Reynal & Hitchcock, New York. 1938.

Mace, C.A. *British Philosophy in the Mid-Century*. George Allen and Unwin Ltd., London. 1957.

Mach, Ernst. *History and Root of the Principle of the Conservation of Energy*. The Open Court Publishing Co., Chicago. 1911.

Mach, Ernst. *The Science of Mechanics*. Sixth Edition. The Open Court Publishing Company, LaSalle. 1960.

Maine, Sir Henry Sumner. *Popular Government: Four Essays*. Henry Holt and Company, New York. 1886.

Manin, Yu. I. *A Course in Mathematical Logic*. Springer-Verlag, New York. 1977.

Mannheim, Karl. *Essays on the Sociology of Knowledge*. Oxford University Press, New York. 1952.

March, Robert H. *Physics for Poets*. McGraw-Hill Book Company, New York. 1970.

Maslow, A.H. *Motivation and Personality*. Harper & Row, Publishers, New York. 1954.

Masters, William H. 'Two Sex Researchers on the Firing Line' in *Life*. June 24, 1966.

Maxwell, James Clerk. *The Collected Papers of James Clerk Maxwell*. Volume 2. At the University Press, Cambridge. No date.

Mayer, J.R. *Mechanik der Wärme*. Stuttgart. 1867.

Mayo, William J. 'Contributions of Pure Science to Progressive Medicine' in *The Journal of the American Medical Association*. Volume 84, Number 20. May 16, 1925.

Mayr, Ernst. *Toward a New Philosophy of Biology*. Harvard University Press, Cambridge. 1988.

McCarthy, Mary. *On the Contrary*. Farrar, Straus and Cudahy, New York. 1961.

McEvoy, John G. 'Electricity, Knowledge, and the Nature of Progress in Priestley's Thought' in *The British Journal of History of Science*. Volume XII, Number 40. March 1979.

McKenzie, John L. *The Two-Edged Sword*. The Bruce Publishing Company, Milwaukee. 1968.

McKuen, Rod. *Listen to the Warm Poems*. Random House, New York. 1967.

Mead, Margaret. *Coming of Age in Samoa*. William Morrow & Company, New York. 1928.

Medawar, Peter. *Advice to a Young Scientist*. Basic Books, Inc., Publishers, New York. 1979.

Medawar, Peter. *Induction and Intuition in Scientific Thought*. American Philosophical Society, Philadelphia. 1969.

Medawar, Peter. *Pluto's Republic*. Oxford University Press, Oxford. 1982.

Medawar, Peter. *The Art of the Soluble*. Methuen & Co., Ltd., London. 1967.

Medawar, Peter. *The Future of Man*. Methuen and Co., Ltd., London. 1960.

Medawar, Peter. *The Hope of Progress*. Anchor Books, Garden City. 1973.

Mellanby, Kenneth. 'Disorganisation of Scientific Research' in *New Scientist*. Volume 59, Number 860. August 23, 1973.

Melville, Herman. *White Jacket*. Oxford University Press, London. 1966.

Mencken, H.L. *Minority Report*. Alfred A. Knopf, New York. 1956.

Mencken, H.L. *Prejudices: Second Series*. Alfred A. Knopf, New York. 1920.

Mercier, André. 'Fifty Years of the Theory of Relativity' in *Nature*. Volume 175. May 28, 1955.

Meyer, Agnes E. *Education for a New Morality*. The Macmillan Company, New York. 1957.

Michelson, A.A. *Light Waves and Their Uses*. The University of Chicago Press, Chicago. 1903.

Middendorf, W.H. and Brown, G.T., Jr. 'Orderly Creative Inventing' in *Electrical Engineering*. October 1957.

Midgley, Mary. 'Can Science Save Its Soul?' in *New Scientist*. August 1, 1992.

Millikan, Robert A. 'The Relation of Science to Industry' in *Science*. Volume LXIX, Number 1776. January 11, 1929.

Millikan, Robert A. *Science and The New Civilization*. Charles Scribner's Sons, New York. 1930.

Milne, E.A. *Modern Cosmology and the Christian Idea of God*. At the Clarendon Press, Oxford. 1952.

Milton, John. 'Paradise Lost' in *Great Books of the Western World*. Volume 32. Encyclopaedia Britannica, Inc., Chicago. 1952.

Montessori, Maria. *The Montessori Method*. Schocker Books, New York. 1964.

Moore, John A. 'Science as a Way of Knowing' in *American Zoologist*. Volume 24, Number 2. 1984.

Moore, John A. *Science as a Way of Knowing*. Harvard University Press, Cambridge. 1993.

More, Louis T. *The Dogma of Evolution*. Princeton University Press, Princeton. 1925.

More, Louis T. *The Limitations of Science*. Henry Holt and Company, New York. 1915.

Morgan, Robin. *Sisterhood is Powerful*. Vintage Books, New York. 1970.

Moszkowski, Alexander. *Conversations with Einstein*. Horizon Press, New York. 1970.

Mozans, H.J. *Women in Science*. University of Notre Dame Press, Notre Dame. 1991.

Muller, Herbert J. *Science & Criticism*. George Braziller, Inc., New York. 1956.

Mumford, Lewis. *Technics and Civilization*. Harcourt, Brace and Company, New York. 1934.

Mumford, Lewis. *The Conduct of Life*. Harcourt, Brace and Company, New York. 1951.

Murrow, Edward R. *This I Believe: 2*. Simon and Schuster, New York. 1954.

Mydral, Gunnar. *Objectivity in Social Research*. Pantheon Books, New York. 1969.

Nahm, Milton C. *John Wilson's The Cheats*. Basil Blackwell, Oxford. 1935.

Nathan, Otto and Norden, Heinz. *Einstein on Peace*. Avenel Books, New York. 1981.

National Academy of Sciences. *Basic Research and National Goals: A Report to the Committee on Science and Astronautics*. March 1965.

Newman, James R. *The World of Mathematics*. Volume Four. Simon and Schuster, New York. 1956.

Newton, Isaac. 'Mathematical Principles of Natural Philosophy' in *Great Books of the Western World*. Volume 34. Encyclopaedia Britannica, Inc., Chicago. 1952.

Newton, Isaac. 'Optics' in *Great Books of the Western World*. Volume 34. Encyclopaedia Britannica, Inc., Chicago. 1952.

Nordmann, Charles. *Einstein and the Universe*. T. Fisher Unwin Ltd., London. 1922.

Norton, Robert. *The Gunner*. Printed by A.M. for Humphrey Robinson. London. 1628.

Obler, Paul C. and Estrin, Herman A. *The New Scientist: Essays on the Methods and Value of Modern Science*. Doubleday & Company, Inc., Garden City. 1962.

Olson, Sigurd F. *Reflections From the North Country*. Alfred A. Knopf, New York. 1976.

O'Neill, Eugene. *Strange Interlude*. Boni & Liveright, New York. 1928.

Oparin, A.I. *Life: Its Nature, Origin and Development*. Academic Press Inc., Publishers, New York. 1962.

Oppenheimer, J. Robert. 'With Oppenheimer on an Autumn Day' in *Look*. Volume 30, Number 26. December 27, 1966.

Oppenheimer, J. Robert. *Foundations for World Order*. The University of Denver Press, Denver. 1949.

Oppenheimer, J. Robert. *Science and the Common Understanding*. Simon and Schuster, New York. 1966.

Oppenheimer, J. Robert. *The Open Mind*. Simon and Schuster, New York. 1955.

Osler, William. *Aequanimitas*. The Blakiston Company, Philadelphia. 1932.

Osler, William. *Evolution of Modern Medicine*. Yale University Press, New Haven. 1921.

Osler, William. *Science and Immortality*. Houghton, Mifflin and Company, Boston. 1904.

Overhage, Carl F.J. 'Science Libraries: Prospects and Problems' in *Science*. Volume 155, Number 3764. February 1967.

Pagels, Heinz R. *The Dreams of Reason: The Computer and the Rise of the Sciences of Complexity*. Simon and Schuster, New York. 1988

Paley, William. *Natural Theology*. American Tract Society, New York. 185?.

Pallister, William. *Poems of Science*. Playford Press, New York. 1931.

Panunzio, Constantine. *Major Social Institutions*. The Macmillan Company, New York. 1947.

Parton, H.N. *Science is Human*. University of Otago Press, Dunedin. 1972.

Pascal, Blaise. 'Pensées' in *Great Books of the Western World*. Volume 33. Encyclopaedia Britannica, Inc., Chicago. 1952.

Paton, H.J. *The Modern Predicament*. George Allen & Unwin Ltd., London. 1955.

Paulos, John Allen. *A Mathematician Reads the Newspaper*. Basic Books, New York. 1995.

Pavlov, Ivan P. *Lectures on Conditioned Reflexes*. Translated by W. Horsely Gantt. Liveright Publishing Corporation, New York. 1928.

Payne-Gaposchkin, Cecilia. *An Autobiography and Other Recollections*. Cambridge University Press, Cambridge. 1984.

Peabody, Francis. 'The Care of the Patient' in *The Journal of the American Medical Association*. Volume 88, Number 12. March 19, 1927.

Pearson, Karl. *The Grammar of Science*. J.M. Dent & Sons, London. 1937.

Peattie, Donald Culross. *Flowering Earth*. G.P. Putnam's Son, New York. 1939.

Peirce, Charles Sanders. 'The Architecture of Theories' in *The Monist*. Volume I, Number 2. January 1891.

Peirce, Charles Sanders. *Chance, Love, and Logic*. Barnes & Noble, Inc., New York. 1968.

Peirce, Charles Sanders. *The Collected Works of Charles Sanders Peirce*. Volume 5. Harvard University Press, Cambridge. 1934.

Penfield, Wilder. *The Difficult Art of Giving*. Little, Brown and Company, Boston. 1967.

Perelman, S.J. *Crazy Like a Fox*. Random House, New York. 1944.

Perry, Ralph Barton. *The Present Conflict of Ideals*. Longmans, Green and Co., New York. 1918.

Petit, Jean-Pierre. *Euclid Rules OK?* John Murray, London. 1982.

Pindar. *The Extant Odes of Pindar*. Macmillan and Co., London. 1888.

Pirsig, Robert M. *Zen and the Art of Motorcycle Maintenance*. Bantam Books, Toronto. 1975.

Pizan, Christine de. *The Book of the City of Ladies*. Translated by Earl Jeffery Richards. Persea Books, New York. 1982.

Planck, Max. *Scientific Autobiography and Other Papers*. Translated by Frank Gaynor. Philosophical Library, New York. 1949.

Planck, Max. *The Philosophy of Physics*. W.W. Norton & Company, Inc., New York. 1963.

Planck, Max. *Where Is Science Going?* W.W. Norton & Company, Inc., New York. 1932.

Poe, Edgar Allen. *Al Aaraaf, Tamerlane and Minor Poems*. Hatch & Dunning, Baltimore. 1829.

Poincaré, Henri. *The Foundations of Science*. The Science Press, New York. 1913.

Polanyi, Michael. 'Faith and Reason' in *Journal of Religion*. Volume XLI, Number 4. October 1961.

Polanyi, Michael. *Science, Faith and Society*. The University of Chicago Press, Chicago. 1946.

Pólya, G. *How to Solve It*. Princeton University Press, Princeton. 1948.

Ponomarev, L.I. *The Quantum Dice*. Institute of Physics Publishing, Bristol. 1993.

Pope, Alexander. *The Works of Alexander Pope*. Volumes II and IV. J. Murray, London. 1871.

Popper, Karl. *Conjectures and Refutations*. Harper & Row, Publishers, New York. 1965.

Popper, Karl. *Objective Knowledge: An Evolutionary Approach*. At the Clarendon Press, Oxford. 1972.

Popper, Karl. *The Logic of Scientific Discovery*. Harper & Row, Publishers, New York. 1968.

Popper, Karl. *Unended Quest*. Open Court Publishing Company, La Salle. 1976.

Porter, Sir George. 'Lest the Edifice of Science Crumble' in *New Scientist*. Volume 111, Number 1524. September 1986.

Porterfield, Austin L. *Creative Factors in Scientific Research*. Duke University Press, Durham. 1941.

Poteat, William Louis. *Can Man Be a Christian To-day?* The University of North Carolina Press, Chapel Hill. 1926.

Praed, Winthrop. *The Poems of Winthrop Mackworth Praed*. Ward, Lock, and Co., London. 1889.

Pratt, C.C. *The Logic of Modern Psychology*. The Macmillan Company, New York. 1939.

Prescott, William Hickling. *History of the Conquest of Mexico*. David McKay, Publishers, Philadelphia. 188?.

President's Science Advisory Committee. *Science, Government, and Information*. The White House. January 10, 1963.

Price, Don K. *The Scientific Estate*. Harvard University Press, Cambridge. 1965.

Priestley, Joseph. *The History and Present State of Electricity*. Johnson Reprint Corporation, New York. 1966.

Prior, Matthew. *Matthew Prior's Literary Works*. Volume I. At the Clarendon Press, Oxford. 1959.

Pritchett, V.S. *The Living Novel & Later Appreciations*. Random House, New York. 1964.

Rabinowitch, Eugene. 'Open Season On Scientists' in *The New Republic*. January 1, 1966.

Raju, P.T. *Idealistic Thought of India*. George Allen & Unwin Ltd., London. 1953.

Rand, Ayn. *Atlas Shrugged*. Random House, New York. 1957.

Rapoport, Anatol. 'Escape from Paradox' in *Scientific American*. Volume 217, Number 1. July 1967.

Rapoport, Anatol. *Operational Philosophy*. Harper & Brothers Publishers, New York. 1953.

Rapoport, Anatol. *Science, Conflict and Society: Readings from Scientific American*. Wilt, Freeman and Company, San Francisco. 1969.

Raven, Charles E. *Science, Religion, and The Future*. At the University Press, Cambridge. 1943.

Ravetz, Jerome R. *Scientific Knowledge and Its Social Problems*. Oxford University Press, New York. 1971.

Redondi, Pietro. *Galileo: Heretic*. Princeton University Press, Princeton. 1987.

Reichen, Charles-Albert. *A History of Astronomy*. Hawthorn Books Inc., New York. 1963.

Reichenbach, Hans. *The Rise of Scientific Philosophy*. University of California Press, Berkeley. 1951.

Renan, Ernest. 'The Nobility of Science' in *Scientific American*. Volume XL, Number 20, New Series. May 17, 1879.

Renan, Ernest. *The Future of Science*. Roberts Brothers, Boston. 1893.

Renan, Ernest. *Life of Jesus*. Little, Brown, and Company, Boston. 1922.

Reston, James, Jr. *Galileo, A Life*. HarperCollins Publishers, New York. 1994.

Reynolds, H.T. *Analysis of Nominal Data*. Number 07-007. Sage Publications, Beverly Hills. 1977.

Richards, I.A. *Principles of Literary Criticism*. Harcourt, Brace & World, Inc., New York. 1925.

Richards, I.A. *Science and Poetry*. Kegan Paul, Trench, Trubner, & Co., London. 1926.

Richet, Charles. *The Natural History of a Savant*. Translated by Sir Oliver Lodge. J.M. Dent & Sons Limited, London. 1927.

Rodley, Chris. *Cronenberg on Cronenberg*. Faber and Faber, London. 1992.

Roe, Anne. *The Making of a Scientist*. Dodd, Mead & Company, New York. 1953.

Rolland, Romain. *Jean-Christophe in Paris*. Henry Holt and Company, New York. 1911.

Romanoff, Alexis. *Encyclopedia of Thoughts*. Ithaca Heritage Books, Ithaca. 1975.

Ross, Sir Ronald. *Memoirs*. E.P. Dutton and Company, New York. 1923.

Rossman, Joseph. *Industrial Creativity: The Psychology of the Inventor*. University Books, New Hyde Park. 1964.

Roszak, Theodore. *The Making of a Counter Culture*. University of California Press, Berkeley. 1995.

Rothschild, Lord Nathaniel Mayer. *A Framework for Government Research and Development*. Her Majesty's Stationary Office, London. 1971.

Rubin, Harry. 'Does Somatic Mutation Cause Most Cancers?' in *Journal of the National Cancer Institute*. Volume 64, Number 5. May 1980.

Ruse, Michael. 'Response to the Commentary: *Pro Judice*' in *Science, Technology & Human Values*. Volume 7, Number 41. Fall 1982.

Russell, Bertrand. *History of Western Philosophy*. George Allen & Unwin Ltd., London. 1961.

Russell, Bertrand. *Human Knowledge: Its Scope and Limits*. George Allen and Unwin Ltd., London. 1948.

Russell, Bertrand. *Our Knowledge of the External World*. George Allen & Unwin Ltd., London. 1926.

Russell, Bertrand. *Religion and Science*. Oxford University Press, London. 1935.

Russell, Bertrand. *The Conquest of Happiness*. Garden City Publishing Company, Inc., Garden City. 1933.

Russell, Bertrand. *The Problems of Philosophy*. Oxford University Press, London. 1959.

Russell, Bertrand. *The Scientific Outlook*. George Allen & Unwin Ltd., London. 1931.

Russell, Bertrand. *Unpopular Essays*. Simon and Schuster, New York. 1950.

Sa'di. *The Gulistān*. Translated by Edward B. Eastwick. Trubner & Co., London. 1880.

Safonov, V. *Courage*. Foreign Languages Publishing House, Moscow. 1953.

Sagan, Carl. *The Demon-Haunted World*. Random House, New York. 1995.

Saint Augustine. *De Magistro*. Hackett Publishing House, Indianapolis. 1995.

Santayana, George. *The Life of Reason*. Charles Scribner's Sons, New York. 1954.

Santayana, George. *The Sense of Beauty*. Charles Scribner's Sons, New York. 1908.

Sarton, George. *History of Science and the New Humanism*. Henry Holt and Company, New York. 1931.

Sarton, George. *Introduction to the History of Science*. Volume I. The Williams & Wilkins Company, Baltimore. 1927.

Sawyer, W.W. *Prelude to Mathematics*. Penguin Books, Ltd., Harmondsworth. 1960.

Sayers, Dorothy L. *Gaudy Night*. Harper & Row, Publishers, New York. 1986.

Schild, A. 'On the Matter of Freedom: The University and the Physical Sciences' in *Canadian Association of University Teachers*. Volume 11, Number 4. 1963.

Schlegel, Friedrich. *Philosophical Fragments*. Translated by Peter Firchow. University of Minnesota Press, Minneapolis. 1991.

Schneer, Cecil J. *Mind and Matter*. Grove Press, Inc., New York. 1969.

Schumacher, E.F. *Small is Beautiful*. Harper & Row, Publishers, New York. 1973.

Schweizer, Karl W. *Herbert Butterfield: Essays on the History of Science*. The Edwin Mellen Press, Lewiston. 1998.

Seifriz, William. 'A New University' in *Science*. Volume 120, Number 3107. July 16, 1954.

Selye, Hans. *From Dream to Discovery*. McGraw-Hill Book Company, New York. 1964.

Serres, Michel. *Hermes: Literature, Science, Philosophy*. The Johns Hopkins University Press, Baltimore. 1982.

Shakespeare, William. 'Julius Caesar' in *Great Books of the Western World*. Volume 26. Encyclopaedia Britannica, Inc., Chicago. 1952.

Shakespeare, William. 'Othello, the Moor of Venice' in *Great Books of the Western World*. Volume 27. Encyclopaedia Britannica, Inc., Chicago. 1952.

Shakespeare, William. 'The Merchant of Venice' in *Great Books of the Western World*. Volume 26. Encyclopaedia Britannica, Inc., Chicago. 1952.

Shapley, Harlow, Rapport, Samuel and Wright, Helen. *A Treasury of Science*. Harper & Row, Publishers, New York. 1965.

Shaw, Bernard. *Back to Methuselah*. Brentano's, New York. 1927.

Shaw, Bernard. *The Doctor's Dilemma*. Brentano's, New York. 1911.

Shepherd, Linda Jean. *Lifting the Veil*. Shambhala, Boston. 1993.

Shermer, Michael. *Why People Believe Weird Things*. W.H. Freeman and Company, New York. 1997.

Siegel, Eli. *Damned Welcome*. Definition Press, New York. 1972.

Sigma Xi. *Honor in Science*. Sigma Xi, The Scientific Research Society, Research Triangle Park. 1991.

Silver, Brian L. *The Ascent of Science*. A Solomon Press Book, New York. 1998.

Singer, Charles. *Studies in the History and Method of Science*. Volume I. William Dawson & Sons, Ltd., London. 1955.

Skinner, Quentin. *The Return of Grand Theory in the Human Sciences*. Cambridge University Press, Cambridge. 1985.

Slater, Robert. *Portraits in Silicon*. The MIT Press, Cambridge. 1987.

Slobodkin, Lawrence B. *Simplicity and Complexity in Games of the Intellect*. Harvard University Press, Cambridge. 1992.

Smith, Adam. *The Wealth of Nations*. The Modern Library, New York. 1937.

Smith, David. 'The Private Thoughts of David Smith' in *Vogue*. November 15, 1968.

Smith, Homer W. *From Fish to Philosopher*. Little, Brown and Company, Boston. 1953.

Smith, Theobald. 'Letter from Dr. Theobald Smith' in *Journal of Bacteriology*. Volume XXVII, Number 1. January 1934.

Smith, Theobald. *New York Medical Journal*. Volume lii. 1890.

Snow, C.P. *The Two Cultures: and A Second Look*. The New American Library, New York. 1964.

Solovine, Maurice. *Lettres à Maurice Solovine*. Gauthier-Villars, Paris. 1956.

Sontag, Susan. *Selected Writings*. University of California Press, Berkeley. 1988.

Spark, Muriel. *The Prime of Miss Jean Brodie*. J.B. Lippincott Company, Philadelphia. 1961.

Spencer, Herbert. *Education*. D. Appleton and Company, New York. 1920.

Spencer, Herbert. *The Data of Ethics*. D. Appleton and Company, New York. 1884.

Spencer-Brown, G. *Laws of Form*. George Allen and Unwin Ltd., London. 1969.

Sperry, Roger. 'The New Mentalist Paradigm and Ultimate Concern' in *Perspectives in Biology and Medicine*. Volume 29, Number 3, Part I. Spring 1986.

Sprat, Thomas. *The History of the Royal Society of London for the Improving of Natural Knowledge*. Printed by T.R. for J. Martyn. London. 1667.

Stace, W.T. *Religion and the Modern Mind*. J.B. Lippincott Company, Philadelphia. 1952.

Stackman, Elvin. 'U.S. Science Holds Its Biggest Powwow' in *Life*. January 9, 1950.

Starling, E.H. 'Discovery and Research' in *Nature*. Volume 113, Number 2843, April 1924.

Sterne, Laurence. *Tristram Shandy*. The Modern Library, New York. 1950.

Stevenson, Robert Louis. *Virginibus Puerisque*. Charles Scribner's Sons, New York. 1887.

Stewart, Ian. *Does God Play Dice?* Basil Blackwell Inc., Cambridge. 1990.

Streeter, Burnett Hillman. *Reality*. The Macmillan Company, New York. 1926.

Strutt, Robert John. *Life of John William Strutt: Third Baron Rayleigh*. The University of Wisconsin Press, Madison. 1968.

Sullivan, J.W.N. 'The Justification of the Scientific Method' in *The Athenaeum*. Number 4644. May 2, 1919.

Sullivan, J.W.N. *Aspects of Science*. Jonathan Cape & Harrison Smith, London. 1923.

Sullivan, J.W.N. *Beethoven, His Spiritual Development*. Alfred A. Knopf, New York. 1964.

Sullivan, J.W.N. *The Limitations of Science*. The New American Library, New York.

Szent-Györgyi, Albert. 'On Scientific Creativity' in *Perspectives in Biology and Medicine*. Volume V, Number 2. Winter 1962.

Szent-Györgyi, Albert. 'Teaching and the Expanding Knowledge' in *Science*. Volume 146, Number 3649. December 4, 1964.

Szent-Györgyi, Albert. *Bioenergetics*. Academic Press Inc., New York. 1957.

Tansley, A.G. 'Classification of Vegetation and the Concept of Development' in *Journal of Ecology*. Volume 8. 1920.

Tate, Allen. 'Critical Responsibility' in *The New Republic*. Volume 51, Number 663. August 17, 1927.

Temple, Frederick. *The Relations between Religion and Science*. Macmillan and Co., New York. 1884.

Temple, G. *Turning Points in Physics*. North-Holland Publishing Company, Amsterdam. 1958.

Tennant, F.R. *Philosophical Theology*. At the University Press, Cambridge. 1956.

Tennyson, Alfred. *The Complete Poetical Works of Tennyson*. Houghton Mifflin Company, Boston. 1898.

Thomas, Lewis. *The Lives of a Cell*. The Viking Press, New York. 1974.

Thompson, Silvanus P. *The Life of William Thomson, Baron Kelvin of Largs*. Volume 2. Macmillan, London. 1910.

Thomson, Sir George. *The Inspiration of Science*. Oxford University Press, London. 1961.

Thomson, J.A. *Introduction to Science*. Henry Holt & Co., New York. 1924.

Thomson, J.A. *The Outline of Science*. G.P. Putnam's Sons, New York. 1922.

Thomson, J.J. *The Life of Sir J.J. Thomson*. At the University Press, Cambridge. 1943.

Thomson, Linda Price. *The Common but Less Frequent Loon and Other Essays*. Yale University Press, New Haven. 1993.

Thomson, William and Tait, P.G. *Treatise on Natural Philosophy*. At the University Press, Cambridge. 1923.

Thoreau, Henry. *The Journal of Henry D. Thoreau*. Volume II. Houghton Mifflin Company, Boston. 1949.

Thoreau, Henry. *Walden*. Published for the Classics Club by W.J. Black, New York. 1942.

Thurber, James. *Collecting Himself*. Harper & Row, Publishers, New York. 1989.

Titchener, Edward Bradford. *Systematic Psychology*. Cornell University Press, Ithaca. 1972.

Todhunter, Isaac. *William Whewell*. Volume I. Macmillan and Co., London. 1876.

Tolstoy, Leo. 'War and Peace' in *Great Books of the Western World*. Volume 51. Encyclopaedia Britannica, Inc., Chicago. 1952.

Tolstoy, Leo. *What is Religion?* Thomas Y. Crowell & Company. New York. 1902.

Toulmin, Stephen E. *The Philosophy of Science*. Hutchinson University Library, London. 1962.

Toynbee, Arnold. *Toynbee's Industrial Revolution*. Augustus M. Kelley, Publishers, New York. 1968.

Trilling, Lionel. *The Liberal Imagination*. Doubleday & Company, Inc., Garden City. 1950.

Trotter, Wilfred. 'Observation and Experiment and Their Use in the Medical Sciences' in *British Medical Journal*. July 26, 1930.

Trotter, Wilfred. 'The Commemoration of Great Men' in *British Medical Journal*. February 20, 1932.

Tupper, Martin. *Proverbial Philosophy*. Wiley & Putnam, New York. 1848.

Twain, Mark. *Is Shakespeare Dead?* Oxford University Press, New York. 1996.

Twain, Mark. *Personal Recollections of Joan of Arc*. Harper & Brothers, Publishers, New York. 1898.

Twain, Mark. *Pudd'nhead Wilson*. Cardavon Press, Inc., Avon. 1974.

Twain, Mark. *The Diaries of Adam and Eve*. Oxford University Press, New York. 1996.

Twain, Mark. *The Innocents Abroad*. Oxford University Press, New York. 1996.

Twain, Mark. *The Complete Works of Mark Twain*. Volume 14. Harper & Brothers, New York. 1911.

Twain, Mark. *What is Man? and Other Essays*. Harper & Brothers Publishers, New York. 1917.

Tyndall, John. *Fragments of Science*. D. Appleton and Company, New York. 1883.

Unamuno, Miguel de. *Essays and Soliloquies*. Translated by J.E. Crawford Flitch. George G. Harrap & Company, London. 1925.

Unamuno, Miguel de. *The Tragic Sense of Life*. Macmillan and Co., Limited, London. 1921.

United Nations. *Women and Science*. United Nations. July 1979.

Vaihinger, H. *The Philosophy of 'As If'*. Routledge & Kegan Paul Ltd., London. 1935.

Valéry, Paul. *The Collected Works of Paul Valéry*. Edited by Jackson Mathews. Volume 14. Princeton University Press, Princeton. No date.

van Fraassen, Bas C. *The Scientific Image*. Clarendon Press, Oxford. 1980.

Vare, Ethlie Ann and Ptacek, Greg. *Mothers of Invention*. William Morrow and Company, Inc., New York. 1988.

Virchow, Rudolf. *Cellular Pathology*. Dover Publications, Inc., New York. 1971.

Virchow, Rudolf. *Disease, Life, and Man*. Translated by Lelland J. Rather. Stanford University Press, Stanford. 1958.

Waddington, C.H. *Science and Ethics*. George Allen & Unwin, Ltd., London. 1944.

Waddington, C.H. *The Nature of Life*. Harper & Row, Publishers, New York. 1966.

Waddington, C.H. *The Scientific Attitude*. Penguin Books, New York. 1941.

Walgate, Robert. 'Breaking Through the Disenchantment' in *New Scientist*. September 18, 1975.

Wallace, Henry A. 'Scientists in an Unscientific Society' in *Scientific Monthly*. Volume 150. 1934.

Wang, H. 'The Formal and the Intuitive in the Biological Sciences' in *Perspectives in Biology and Medicine*. Volume 27, Number 4, 1984.

Watson, David Lindsay. *Scientists are Human*. Watts & Co., London. 1938.

Watson, James D. *The Double Helix*. Atheneum, New York. 1968.

Weaver, Jefferson Hane. *The World of Physics*. Volume II. Simon and Schuster, New York. 1987.

Weber, Max. *Essays in Sociology*. Translated by H.H. Gerth and C. Wright Mills. A Galaxy Book, New York. 1958.

Weber, Renée. *Dialogues with Scientists and Sages: The Search for Unity*. Routledge and Kegan Paul, London. 1986.

Weil, Simone. *Gravity and Grace*. G.P. Putnam's Sons, New York. 1952.

Weil, Simone. *On Science, Necessity, and the Love of God*. Translated by Richard Rees. Oxford University Press, London. 1968.

Weil, Simone. *Oppression and Liberty*. Translated by Arthur Wills and John Petrie. Routledge and Kegan Paul Ltd., London. 1958.

Weil, Simone. *The Need for Roots*. Translated by Arthur Wills. The Beacon Press, Boston. 1952.

Weil, Simone. *The Simone Weil Reader*. David McKay Company, Inc., New York. 1977.

Weinberg, Alvin M. *Reflections on Big Science*. The M.I.T. Press, Cambridge. 1967.

Weinberg, S. 'Newtonianism, Reductionism and the Art of Congressional Testimony' in *Nature*. Volume 330. December 3, 1987.

Weismann, August. *The Evolution Theory*. Volume I. Translated by J. Arthur Thomson. Edward Arnold, London. 1904.

Weiss, Paul A. *Within the Gates of Science and Beyond*. Hafner Publishing Company, New York. 1971.

Weisskopf, Victor. 'The Significance of Science' in *Science*. Volume 176, Number 4031. April 14, 1972.

Weisskopf, Victor. *The Joy of Insight*. Basic Books, New York. 1991.

Weisskopf, Victor F. *The Privilege of Being a Physicist*. W.H. Freeman and Company, New York. 1989.

Wells, H.G. *New Worlds for Old*. The Macmillan Company, New York. 1908.

Westbrock, Peter. 'The Oceans Inside Us' in *The Times Higher Education Supplement* (London). November 3, 1995.

Wheeler, Edgar C. 'Makers of Lightning' in *The World's Work*. January 1927.

Whetham, William Cecil Dampier. *The Recent Development of Physical Science*. P. Blakiston's Sons & Co., Philadelphia. 1904.

Whewell, William. *History of the Inductive Sciences, From the Earliest to the Present Time*. J.W. Parker, London. 1837.

Whewell, William. *The Philosophy of the Inductive Sciences Founded upon Their History*. Volume the Second. John W. Parker, London. 1847.

Whitehead, Alfred North. *Adventures of Ideas*. The Macmillan Company, New York. 1956.

Whitehead, Alfred North. *An Introduction to Mathematics*. Oxford University Press, New York. 1948.

Whitehead, Alfred North. *Modes of Thought*. The Macmillan Company, New York. 1957.

Whitehead, Alfred North. *Nature and Life*. Greenwood Press, Publishers, New York. 1934.

Whitehead, Alfred North. *Religion in the Making*. At the University Press, Cambridge. 1927.

Whitehead, Alfred North. *Science and the Modern World*. The Free Press, New York. 1967.

Whitehead, Alfred North. *The Aims of Education*. The Macmillan Company, New York. 1929.

Whitehead, Alfred North. *The Concept of Nature*. At the University Press, Cambridge. 1930.

Whitehead, Alfred North. *The Function of Reason*. Princeton University Press, Princeton. 1929.

Whitehead, Alfred North. *The Organisation of Thought*. Williams and Norgate, London. 1917.

Whitney, W.R. 'The Stimulation of Research in Pure Science which Has Resulted from the Needs of Engineers and of Industry' in *Science*. Volume LXV, Number 1862. March 25, 1927.

Whitrow, G.J. *Einstein: The Man and His Achievement*. British Broadcasting Corporation, London. 1967.

Whyte, Lancelot Law. *Accent on Form*. Harper & Brothers Publishers, New York. 1954.

Whyte, Lancelot Law. *The Unconscious before Freud*. Basic Books, Inc., New York. 1960.

Wiener, Norbert. 'A Rebellious Scientist After Two Years' in *Bulletin of the Atomic Scientists*. Volume 4, Number 11. November 1948.

Wiener, Norbert. *I Am a Mathematician*. The M.I.T. Press, Cambridge. 1958.

Wilde, Oscar. *An Ideal Husband*. Leonard Smithers and Co., London. 1899.

Wilder, Thornton. 'Thornton Wilder at 65 Looks Ahead—And Back' in *New York Times Magazine*. April 15, 1962.

Wilkins, John. *The Discovery of a World in the Moone*. Printed by E.G. for Michael Sparke and Edward Forrest. London. 1638.

Williams, L. Pearce. 'Letter to the Editor' in *Scientific American*. Volume 214, Number 6. June 1966.

Willstätter, Richard. *From My Life*. W.A. Benjamin, Inc., New York. 1965.

Wilson, Edmund. *Letters On Literature and Politics 1912–1972*. Farrar, Straus and Giroux, New York. 1977.

Wilson, E. Bright. *An Introduction to Scientific Research*. McGraw-Hill Book Company, Inc., New York. 1952.

Wilson, Edwin B. 'The Statistical Significance of Experimental Data' in *Science*. Volume 58, Number 1493. August 1923.

Wilson, Edward O. *Biophilia*. Harvard University Press, Cambridge. 1984.

Wisdom, John Oulton. *Foundations of Inference in Natural Science*. Methuen & Co., Ltd., London. 1952.

Wittgenstein, Ludwig. *Culture and Value*. Translated by Peter Winch. The University of Chicago Press, Chicago. 1980.

Wittgenstein, Ludwig. *Tractatus Logico-Philosophicus*. Routledge & Kegan Paul, Ltd., London. 1922.

Wolfle, Dael. *Symposium on Basic Research*. American Association for the Advancement of Science, Washington, D.C. 1959.

Wolpert, Lewis. *The Unnatural Nature of Science*. Harvard University Press, Cambridge. 1993.

Wordsworth, William. *The Complete Poetical Works of William Wordsworth*. Macmillan and Co., London. 1891.

Yates, Frances A. *Giordano Bruno and the Hermetic Tradition*. The University of Chicago Press, Chicago. 1964.

Yentsch, Clarice M. and Sindermann, Carl J. *The Woman Scientist*. Plenum Press, New York. 1992.

Zemanian, Armen H. 'Appropriate Proof Techniques' in *The Physics Teacher*. Volume 32, Number 5. May 1994.

Ziman, J.M. 'Information, Communication, Knowledge' in *Nature*. Volume 224, Number 5217. October 25, 1969.

Ziman, John. *Knowing Everything About Nothing*. Cambridge University Press, Cambridge. 1987.

Ziman, J.M. *Public Knowledge*. At the University Press, Cambridge. 1968.

Ziman, John. *Reliable Knowledge*. Cambridge University Press, Cambridge. 1978.

Zinsser, Hans. 'Untheological Reflections' in *The Atlantic Monthly*. July 1929.

Zinsser, Hans. *As I Remember Him*. Little, Brown and Company, Boston. 1940.

Zworykin, V.K. 'You Can Write it Down . . .' in *The American Magazine*. August 1949.

PERMISSIONS

Grateful acknowledgment is made to the following for their kind permission to reprint copyright material. Every effort has been made to trace copyright ownership but if, inadvertently, any mistake or omission has occurred, full apologies are herewith tendered.

Full references to authors and the titles of their works are given under the appropriate quotation.

A COURSE IN MATHEMATICAL LOGIC by Yu. I. Manin. Copyright 1977. Reprinted by permission of the publisher Springer-Verlag, New York.

A MATHEMATICIAN READS THE NEWSPAPER by John Allen Paulos. Copyright 1995. Reprinted by permission of the publisher Basic Books, Inc., New York.

A SENSE OF THE FUTURE by J. Bronowski. Reprinted by permission of the publisher The MIT Press, Cambridge.

A WORLD ON PAPER by Enrico Bellone. Reprinted by permission of the publisher The MIT Press, Cambridge.

ADVICE TO A YOUNG SCIENTIST by Peter Medawar. Copyright 1979. Reprinted by permission of the publisher Basic Books, Inc., New York.

AN INTRODUCTION TO MATHEMATICS by Alfred North Whitehead. Copyright 1948. Reprinted by permission of the publisher Oxford University Press.

BRIGHT AIR, BRILLIANT FIRE by Gerald M. Edelman. Copyright 1992. Reprinted by permission of the publisher Basic Books, Inc., New York.

COSMIC DISCOVERY by Martin Harwit. Copyright 1981. Reprinted by permission of the publisher Basic Books, Inc., New York.

CULTURE AND VALUE by Ludwig Wittgenstein. Reprinted by permission of the publisher The University of Chicago Press, Chicago.

ESSAYS IN POSITIVE ECONOMICS by Milton Friedman. Reprinted by permission of the publisher The University of Chicago Press, Chicago.

ESSAYS ON THE SOCIOLOGY OF KNOWLEDGE by Karl Mannheim. Copyright 1952. Reprinted by permission of the publisher Oxford University Press.

EXPERIENCING SCIENCE by Jeremy Bernstein. Copyright 1978. Reprinted by permission of the publisher Basic Books, Inc., New York.

EXPLAINING SCIENCE by Ronald N. Giere. Reprinted by permission of the publisher The University of Chicago Press, Chicago.

FOUNDATIONS OF MODERN PHYSICAL SCIENCE by Gerald Holton and Duane H.D. Roller. Copyright 1958. Reprinted by permission of the publisher Addison-Wesley Publishing Company, Reading.

FROM MAX WEBER by H.H. Gerth and C. Wright Mills. Copyright 1946. Reprinted by permission of the publisher Oxford University Press.

GIORDANO BRUNO AND THE HERMETIC TRADITION by Frances A. Yates. Reprinted by permission of the publisher The University of Chicago Press, Chicago.

INFORMATION, TECHNOLOGY, AND CIVILIZATION by Hiroshi Inose and John R. Pierce. Reprinted by permission of the publisher W.H. Freeman and Company, New York.

IS SHAKESPEARE DEAD? by Mark Twain. Copyright 1996. Reprinted by permission of the publisher Oxford University Press.

KNOWLEDGE by Keith Lehrer. Reprinted by permission of the author.

KNOWLEDGE & THE FLOW OF INFORMATION by Fred Dretske. Reprinted by permission of the publisher The MIT Press, Cambridge.

LIGHT WAVES AND THEIR USES by A.A. Michelson. Reprinted by permission of the publisher The University of Chicago Press, Chicago.

LINGUISTIC ASPECTS OF SCIENCE by Leonard Bloomfield. Reprinted by permission of the publisher The University of Chicago Press, Chicago.

MATHEMATICS AND THE SEARCH FOR KNOWLEDGE by Morris Kline. Copyright 1985. Reprinted by permission of the publisher Oxford University Press.

MATHEMATICS IN WESTERN CULTURE by Morris Kline. Copyright 1953. Reprinted by permission of the publisher Oxford University Press.

MIDDLEMARCH by George Eliot. Copyright 1988. Reprinted by permission of the publisher Oxford University Press.

NATURAL PHILOSOPHY OF CAUSE AND CHANCE by Max Born. Copyright 1949. Reprinted by permission of the publisher Oxford University Press.

NATURE'S IMAGINATION by John Cornwell. Copyright 1995. Reprinted by permission of the publisher Oxford University Press.

OBJECTIVE KNOWLEDGE: AN EVOLUTIONARY APPROACH by Karl Popper. Reprinted by permission of Alfred Raymond Mew and Melitta Mew.

ON SCIENCE, NECESSITY, AND THE LOVE OF GOD by Simone Weil. Copyright 1968. Reprinted by permission of the publisher Oxford University Press.

PHYSICS FOR POETS by Robert H. March. Reprinted by permission of the publisher The McGraw-Hill Companies, New York.

PHYSICS IN MY GENERATION by Max Born. Copyright 1969. Reprinted by permission of the publisher Springer-Verlag, New York.

PLUTO'S REPUBLIC by Peter Medewar. Copyright 1982. Reprinted by permission of the publisher Oxford University Press.

PORTRAITS IN SILICON by Robert Slater. Reprinted by permission of the publisher The MIT Press, Cambridge.

RELIGION AND SCIENCE by Bertrand Russell. Copyright 1935. Reprinted by permission of the publisher Oxford University Press.

SCIENCE AND HUMAN LIFE by J.A.V. Butler. Copyright 1957. Reprinted by permission of the publisher Basic Books, Inc., New York.

SCIENCE AND THE PUBLIC MIND by Benjamin Gruenberg. Reprinted by permission of the publisher The McGraw-Hill Companies, New York.

SCIENCE, FAITH AND SOCIETY by Michael Polanyi. Reprinted by permission of the publisher The University of Chicago Press, Chicago.

SCIENTIFIC CHANGE by A.C. Crombie. Copyright 1963. Reprinted by permission of the publisher Basic Books, Inc., New York.

SCIENTIFIC KNOWLEDGE AND ITS SOCIAL PROBLEMS by Jerome R. Ravetz. Copyright 1971. Reprinted by permission of the author and of Oxford University Press.

THE ART OF CONJECTURE by Bertrand de Jouvenal. Copyright 1967. Reprinted by permission of the publisher Basic Books, Inc., New York.

THE ARTFUL UNIVERSE by John D. Barrow. Copyright 1995. Reprinted by permission of the publisher Oxford University Press.

THE COMPLETE WRITINGS OF WILLIAM BLAKE by William Blake. Copyright 1966. Reprinted by permission of the publisher Oxford University Press.

THE CREATIVE LOOP: HOW THE BRAIN MAKES A MIND by Erich Harth. Copyright 1993. Reprinted by permission of the publisher Addison-Wesley Publishing Company, Reading.

THE END OF SCIENCE by John Horgan. Copyright 1996. Reprinted by permission of the publisher Addison-Wesley Publishing Company, Reading.

THE INSPIRATION OF SCIENCE by Sir George Thomson. Copyright 1961. Reprinted by permission of the publisher Oxford University Press.

THE LITERARY WORKS OF LEONARDO DA VINCI by Leonardo da Vinci. Copyright 1939. Reprinted by permission of the publisher Oxford University Press.

THE MISMEASURE OF MAN by Stephen Jay Gould. Copyright 1981 by Stephen Jay Gould. Reprinted by permission of W.W. Norton & Company, Inc.

THE MODES OF THOUGHT by Alfred North Whitehead. Copyright 1938. Reprinted by permission of Simon & Schuster.

THE NEW LEVIATHAN by R.G. Collingwood. Copyright 1942. Reprinted by permission of the publisher Oxford University Press.

THE ORIGINS OF MODERN SCIENCE, 1300–1800 by Herbert Butterfield. Copyright 1957. Reprinted by permission of Simon & Schuster.

THE PRIVILEGE OF BEING A PHYSICIST by Victor F. Weisskopf. Reprinted by permission of the publisher W.H. Freeman and Company, New York.

THE PROBLEMS OF PHILOSOPHY by Bertrand Russell. Copyright 1959. Reprinted by permission of the publisher Oxford University Press.

'The Rationality of Scientific Revolutions' by Karl Popper in Rom Harré (Editor) PROBLEMS OF SCIENTIFIC REVOLUTION. Reprinted by permission of Alfred Raymond Mew and Melitta Mew.

THE SCIENTIFIC REVOLUTION by H. Floris Cohen. Reprinted by permission of the publisher The University of Chicago Press, Chicago.

THE SOCIAL FUNCTION OF SCIENCE by J.D.Bernal. Copyright 1939. Reprinted by permission of the publisher Simon & Schuster.

THE STRUCTURE OF SCIENTIFIC REVOLUTIONS by Thomas S. Kuhn. Reprinted by permission of the publisher The University of Chicago Press, Chicago.

THE UNCONSCIOUS BEFORE FREUD by Lancelot Law Whyte. Copyright 1960. Reprinted by permission of the publisher Basic Books, Inc., New York.

THE WORKS OF THE MIND by Robert B. Heywood. Reprinted by permission of the publisher The University of Chicago Press, Chicago.

TRUTH AND BEAUTY by S. Chandraseckhar. Reprinted by permission of the publisher The University of Chicago Press, Chicago.

UNIVERSITIES by Abraham Flexner. Copyright 1968. Reprinted by permission of the publisher Oxford University Press.

WHAT DO YOU CARE WHAT OTHER PEOPLE THINK?: Further Adventures of a Curious Character by Richard P. Feynman as told to Ralph Leighton. Copyright 1988 by Gweneth Feynman and Ralph Leighton. Reprinted by permission of W.W. Norris & Company, Inc.

WHAT MAD PURSUIT by Francis Crick. Copyright 1988. Reprinted by permission of the publisher Basic Books, Inc., New York.

WHITE JACKET by Herman Melville. Copyright 1966. Reprinted by permission of the publisher Oxford University Press.

SUBJECT BY AUTHOR INDEX

AUTHOR BY SUBJECT INDEX

-C-

-D-

-G-

-I-

-J-

-K-

-L-

-N-

-Q-

-R-

-T-

-U-

-V-

-Y-

-Z-

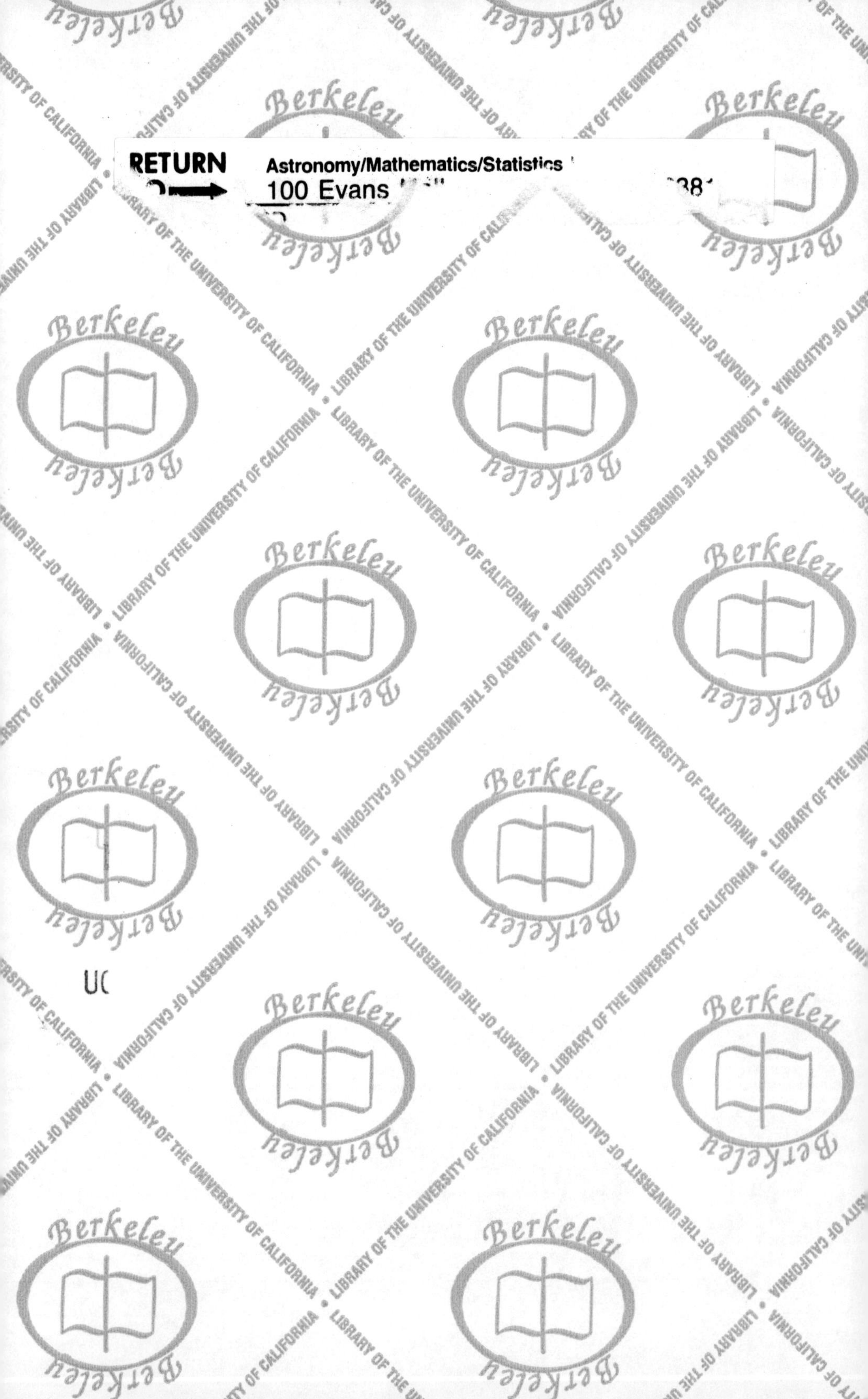